ISBN 978-1-331-42466-6
PIBN 10188310

# BRITISH BIRDS' NESTS

HOW, WHERE, AND WHEN TO
FIND AND IDENTIFY THEM

BY

RICHARD KEARTON, F.Z.S.

ILLUSTRATED FROM PHOTOGRAPHS BY

CHERRY AND RICHARD KEARTON

*WITH COLOURED AND REMBRANDT PLATES*

NEW EDITION REVISED AND ENLARGED



CASSELL AND COMPANY, LIMITED
LONDON, PARIS, NEW YORK, TORONTO AND MELBOURNE.
1908

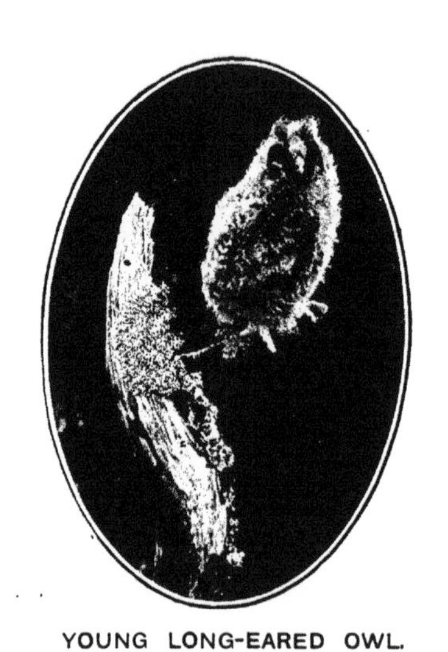

YOUNG LONG-EARED OWL.

# PREFACE

ON the tenth day of April, eighteen hundred and ninety-two, my brother photographed the nest and eggs of a Song Thrush in the neighbourhood of London, and the result appeared to me to be so full of promise that I at once determined to write a book on British Birds' Nests and get him to illustrate it from beginning to end by photographs taken *in situ*. Before mentioning the idea to my publishers, however, I submitted an article on the subject, together with a number of photographs of nests and eggs, to the Editor of an illustrated weekly periodical as a feeler. The prompt acceptance of the contribution, accompanied by a request for more materials of the same character, filled author and illustrator with hope, and we laboured unremittingly spring by spring, without experience or suggestion of any kind on the pictorial side of our work, in the preparation of the original edition of this book, which first saw the light in the autumn of eighteen hundred and ninety-five.

It was the first book of its kind to be illustrated throughout by means of photographs taken direct from Nature and, as a great authority said at the time, " show things as they are and not as they are supposed to be."

When it appeared, Dr. Bowdler Sharpe, of the British Museum, South Kensington, said that it "marked a new era in natural history." I must frankly confess that at the time I regarded these prophetic words in the light of fatherly encouragement from one of the most kind-hearted men in Britain, but a few years have sufficed to prove their absolute truth.

Since the first edition of "British Birds' Nests" was issued, my brother and I have spared neither labour nor money to secure additional photographs not only of nests and eggs, but of the birds themselves. "Our Rarer British Breeding Birds" was published as a supplementary volume in eighteen hundred and ninety-nine, and embodied the result of our further researches up to that time. The present Revised and Enlarged Edition of the major work having destroyed the reason for the further existence of the minor one, we have included the best of the pictures appearing in its pages, together with all the appropriate photographs secured during the last seven years.

It may be of interest to mention that, from first to last, we have travelled by railroad and steamboat alone upwards of thirty thousand miles, and exposed over ten thousand plates in pursuit of bird photography.

The value of the camera where truth and accuracy are a desideratum may easily be seen by noting the distinguishing sexual marks over the eyes of the male and female Red-Necked Phalarope figured on page 277, and comparing the illustration with the coloured plates representing this species in some of the best books ever published on British Birds.

Of the benefit which has accrued to the wild birds of our country by the introduction of this harmless and yet truly sporting method of studying Nature, the following extract from one of many similar letters bears eloquent testimony: "I consider that the birds ought to be extremely grateful to you for inventing bird-photography. I never knew anything that has done so much for their protection during the nesting season as your example. I myself have for several years given up egg collecting entirely, and know many others who have done the same." As further evidence in the same direction I may mention that I had shown to me last spring a Hen Harrier's nest within forty miles of London, and although it was found and photographed by several different naturalists of the new school the old birds carried away two young ones in safety.

It is devoutly to be wished that some of the collectors who are doing so much harm to our rarest breeding birds would either follow this worthy example, or, at any rate, moderate their depredations to the taking of one clutch. When one man can take three or four clutches of eggs belonging to a rare and interesting species, and another shoot, in spite of the law, eight specimens at one shot, of a bird that used to breed regularly in our islands, and would do so again if allowed, the true bird lover is left in despair over the prospects of every species that can be called "rare." Once this unfortunate adjective can be applied to a bird a premium is put upon its skin and eggs. It is a curious kind of morality that will scorn to steal from the individual and yet rob the community without compunction. Wild birds are National property, and no individual has a right to harm one of them without the sanction of the law to do so.

I plead earnestly for our rare breeding birds in danger of extinction. It is far more interesting to any man who can be called an ornithologist and not a mere collector of bric-à-brac to see the living representative of a species soaring majestically over a mountain top than to gaze at its empty egg shells in a cabinet.

Our very sincere thanks are tendered to the many good friends who have placed us under a deep debt of gratitude by the facilities they have most ungrudgingly given us, on all occasions, to secure the photographs scattered up and down the pages of this book. It is unnecessary to say how grateful we shall feel to anyone who may be fortunate enough to find a rare nest, not already figured in this work, for an opportunity to photograph it. We also have to acknowledge our great indebtedness to Messrs. Watkins and Doncaster, Naturalists in the Strand, London, for kindly lending our publishers the specimens from which the plates of eggs were photographed and printed by the three-colour process.

It is, perhaps, hardly necessary to point out, even to the inexperienced ornithologist, that the circumstances governing the distances at which our photographs have been taken rendered it impossible to adjust the relative sizes of the birds and eggs of different species. However, the life size of the eggs in the coloured plates and the average measurements of them and the birds given in the text, will easily prevent any kind of confusion.

The descriptions of the plumage and general appearance of males and females of each species dealt with are, of course, of adult birds in the breeding season.

R. KEARTON.

CATERHAM VALLEY, SURREY,
*January*, 1908.

# LIST OF REMBRANDT PLATES


Young Buzzard . . . . . . . *Frontispiece*

Bullfinch on Nest . . . . . . . *To face p.* 10

Black-Throated Diver on Nest . . . . " " 70

Great Crested Grebe on Nest . . . . " " 136

Young Kites in Nest . . . . . . " " 202

Marsh Warbler and Nest . . . . . " " 466

# LIST OF EGG PLATES

PLATE I.—Rook—Carrion Crow—Jackdaw—Starling—Raven —Jay—Magpie—Hooded Crow—Chough . *To face p.* 28

PLATE II.—Cirl Bunting—Corn Bunting—Crossbill—Snow Bunting—Yellow Hammer—Tree Sparrow—Reed Bunting—Hawfinch—Linnet—House Sparrow—Greenfinch —Bullfinch—Twite—Chaffinch—Lesser Redpole—Goldfinch—Meadow Pipit—Woodlark—Skylark—Rock Pipit . . . . . . . . . . *To face p.* 46

PLATE III.—Blue-Headed Wagtail—Grey Wagtail—Pied Wagtail—White Wagtail—Yellow Wagtail—Bearded Tit—Blue Tit—Coal Tit—Crested Tit—Great Tit—Marsh Tit—Long-Tailed Tit—Tree Creeper—Gold-Crest —Tree Pipit—Woodchat Shrike—Red-Backed Shrike—Siskin—Chiffchaff—Nuthatch—Whitethroat—Lesser Whitethroat—St. Kilda Wren—Common Wren—Dartford Warbler . . . . . . . . *To face p.* 88

PLATE IV.—Wood Warbler—Garden Warbler—Marsh Warbler—Sedge Warbler—Blackcap—Grasshopper Warbler—Willow Warbler—Reed Warbler—Missel Thrush—Ring Ouzel—Blackbird—Stonechat—Song Thrush—Redstart—Wheatear—Robin—Whinchat—Nightingale . . . . . . . . *To face p.* 102

PLATE V.—Hedge Sparrow—Spotted Flycatcher—Pied Flycatcher—Swallow—Cuckoo—Peregrine Falcon—Cuckoo —Hobby—Nightjar—Merlin . . . . *To face p.* 152

PLATE VI.—Osprey—Kestrel—Sparrow-Hawk—Golden Eagle . . . . . . . . . . *To face p.* 184

PLATE VII.—Lapwing—Golden Plover—Ringed Plover—Kentish Plover—Common Dotterel—Kite—Common Buzzard . . . . . . . . *To face p.* 216

PLATE VIII.—Stone Curlew—Ruff—Wood Sandpiper—Red-Necked Phalarope — Common Sandpiper — Oyster-Catcher—Heron . , . . . . . *To face p.* 242

PLATE IX.—Redshank — Greenshank — Woodcock—Dunlin — Common Snipe — Common Curlew — Whimbrel *To face p.* 274

PLATE X.—Herring Gull—Black-Headed Gull—Kittiwake—Great Black-Backed Gull . . . . *To face p.* 308

PLATE XI.—Lesser Black-Backed Gull—Common Gull—Richardson's Skua—Common Skua . . *To face p.* 344

PLATE XII.—Roseate Tern—Sandwich Tern—Common Guillemot — Common Tern — Little Tern — Arctic Tern . . . . . . . . . *To face p.* 370

PLATE XIII.—Red-Throated Diver — Black Guillemot — Puffin—Black-Throated Diver . . . *To face p.* 408

PLATE XIV.—Storm Petrel—Moorhen—Leach's Fork-Tailed Petrel—Razorbill—Corn Crake—Coot—Spotted Crake *To face p.* 440

PLATE XV.—Red-Legged Partridge—Red Grouse—Ptarmigan—Common Partridge—Pheasant—Water Rail—Capercaillie—Quail—Black Grouse . . *To face p* 484

# BRITISH BIRDS' NESTS

## ACCENTOR. *See* HEDGE SPARROW.

## BITTERN.

(*Botaurus stellaris.*)

Order HERODIONES ; Family ARDEIDÆ (HERONS).

HAS now quite ceased to breed in the British Isles, and is only a much persecuted visitor to our shores.

## BLACKBIRD.

(*Turdus merula.*)

Order PASSERES ; Family TURDIDÆ (THRUSHES).

MALE BLACKBIRD.

*Description of Parent Birds.*—Length about ten inches. Bill of medium length, nearly straight, and yellow. Irides yellow. Plumage uniform deep black. Legs and toes brownish-black, claws black.

The female is of a dark rusty-brown colour, lightest and brightest on the throat and breast; bill and feet dusky-brown.

*Situation and Locality.*—Bushes, hedges, ledges of rock, holes in and on projecting "throughs" of dry walls, on banks, in evergreens, against the trunks of trees, on beams in sheds and barns. I have found several built quite on the ground, where a partridge might have been expected to nest, although there were thick old hedgerows not many yards away. On one occasion I discovered a nest inside a thrashing machine standing in an open field, and another in a hole in a tree where a starling would no doubt have bred had the entrance hole been a trifle smaller. I have upon more than one occasion seen a Blackbird's nest built upon an old Thrush's, and *vice versâ*, and occupied nests of the two species touching each other. Common nearly all over the United Kingdom. I have met with it breeding in the Outer Hebrides.

*Materials.*—Small twigs, roots, dry grass, moss, intermixed with clay or mud. Sometimes bits of wool, leaves, fern fronds, and even paper, lined internally with fine dry grass.

*Eggs.*—Four or five, sometimes six. Some authorities say as many as seven and eight; but I have never found a nest with more than six in, although a friend—upon whose word I can place absolute reliance—recently showed me a clutch of eight, and the nest in which he had found them. Of a dull bluish-green, spotted and blotched, and rarely streaked with reddish-brown and grey. They vary considerably, both in regard to ground-colour, shape, size, and markings. Some varieties are covered with small spots, others with such large ones that they very closely resemble the eggs of the Ring Ouzel, whilst a third variety is almost

BLACKBIRDS' NESTS AND EGGS.

spotless. Size about 1.18 by .85 in. (*See* Plate IV.)

*Time.*—March, April, May, June, July, and even as late as August.

*Remarks.*—Resident. Notes: call, *tsissrr, tack, tack;* alarm, a loud, ringing *spink,* or *chink, chink, chink.* Song powerful, and generally delivered at the beginning and end of the day, or during the falling of a warm summer shower, which always seems to delight the heart of this highly esteemed feathered vocalist. Local and other names: Merle, Black Ouzel, Amzel Ouzel. A close sitter. My friend Miss Robinson, of Darlington, tells me in a letter that a Blackbird and a Song Thrush built nests in her garden this spring (1906). The latter laid four eggs, and after a day or two forsook them. The former, instead of laying in her own nest, took possession of that of her neighbour, and commenced the work of incubation, which she carried on until the nest was robbed by boys.

YOUNG FEMALE BLACKBIRD.

## BLACKCAP.

(*Sylvia atricapilla.*)

Order PASSERES ; Family SYLVIIDÆ (WARBLERS).

BLACKCAP ON NEST.

*Description of Parent Birds.* — Length about six inches ; bill of medium length, straight, and dark horn colour. Irides brown. All the upper part of the head black ; nape ash-grey ; back and wing-coverts ash-grey, tinged with brown ; wing and tail-quills brown, bordered with grey ; cheeks, chin, throat, and breast light grey ; belly and under-parts white ; legs and toes lead colour ; claws brown.

The female is larger than the male ; the top of her head is dull rust colour, and her plumage generally more tinged with brown.

*Situation and Locality.*—In brambles, briars, thick hedges, nettles, and gooseberry bushes, in gardens, orchards, thickets, shrubberies, and other suitable places, at heights varying from two to ten or twelve feet from the ground. In all parts of England and Wales, and more sparingly in Scotland and Ireland. It is a very widely distributed bird.

*Materials.*—Fibrous roots, straws, and dead grass, with an inner lining of hair. It is a flimsy

BLACKCAP'S NEST AND EGGS.

structure, sometimes strengthened by wool or spiders' webs.

*Eggs.*—Five or six, very variable, and often difficult to identify, as they closely resemble those of some of the other Warblers. The commonest type is that of a greyish-white underground, suffused with buffish-brown, and spotted, blotched, and marbled with dark brown. Sometimes they are found pale brick-red, marked with a darker tinge of the same colour, and reddish-brown; also faint blue, marked with grey and yellowish-brown. Size about .78 by .58 in. (*See* Plate IV.)

*Time.*—May and June.

*Remarks.*—Migratory, arriving in April and leaving in September, odd specimens remaining till November or December. Notes: alarm, *tack, tack*, or *tec, tec.* Song of great power and freedom. Local and other names: Hay Jack, Hay Chat, Mock Nightingale, Nettle-creeper, Nettle-monger, Blackcap Warbler. Sits very closely. The male bird sometimes takes his share of the task of incubation, and, it is said, relieves the tedium of his task by singing whilst doing so. I have once or twice seen a male bird of this species covering the eggs, but have never yet heard him sing on the nest.

---

## BULLFINCH.

(*Pyrrhula europæa.*)

Order PASSERES; Family FRINGILLIDÆ (FINCHES).

*Description of Parent Birds.*—Length about six inches. Bill short, broad, and thick at the base, and black. Irides dark brown. Round the base of the beak, and all the upper part of the head, black. Nape, back, and lesser wing-coverts grey; greater coverts black, tipped with greyish-white, which forms a bar across the wing-quills, which are dusky. Rump white, upper tail-coverts and tail-quills black. Cheeks, breast, and belly tile-red. Vent and under tail-coverts white. Legs and toes flesh colour; claws brown.

The female has the black on the head, wings, and tail not so intense; her nape and back are greyish-brown, and breast, belly, and under-parts dirty brown.

*Situation and Locality.*—In the lower branches of trees, in evergreens, the tops of high bushes, in thick quick-set hedges and thickets, in suitable localities throughout the British Isles; rarer towards the extreme north of Scotland and in Ireland. The one represented in our illustration on the opposite page was situated in a thick wood, although, according to my experience, the bird is fonder of rough, open, uncultivated land with scattered clumps of thorn bushes for breeding in.

*Materials.*—Small twigs and fibrous roots, interlaced so as to form, as a rule, a broad and flat platform, in the centre of which is the cup-shaped recess lined with fine fibrous roots and sometimes a little wool, hair, or a few feathers.

*Eggs.*—Four to six. Pale greenish-blue, spotted, speckled, and sometimes streaked with dark purplish-brown, and with underlying blotches of brownish-pink. The markings generally form a zone round the large end of the egg. Size about .77 by .57 in. (*See* Plate II.)

*Time.*—April, May, June, and July. I have seen a brood of half-grown young ones, however, in the nest as late as the middle of September.

BULLFINCH'S NEST AND EGGS.

*Remarks.*—Resident. Notes: call, soft, plaintive, and frequently uttered; song, feeble, and low. Male ceases to sing, it is said, as soon as eggs have been laid. However, I have heard an individual which was close to one of my hiding tents in a wood singing in July. Local and other names: Beechfinch, Horsefinch, Pink, Twink, Olph, Nope,

MALE BULLFINCH.

Red Hoop, Alp, Hoop. Sits very closely indeed. This species has benefited more, perhaps, by the Wild Birds' Protection Acts than any other British breeder.

---

## BUNTING, CIRL.

(*Emberiza cirlus.*)

Order PASSERES; Family EMBERIZIDÆ (BUNTINGS).

*Description of Parent Birds.*—Length about six inches. Bill short, conical, and bluish-grey. Irides hazel. A streak of light yellow runs from the base of the upper mandible, over the eye, behind the

BULLFINCH ON NEST.

ear-coverts, and thence forward under the black on the throat; and another from the gape under the eye; crown, behind the eye and under the lower yellow streak, olive-brown streaked with black; the head, nape, and sides of the neck dark olive; back rich chestnut-brown, the margins of the feathers being tinged with olive; wings dusky-black edged with chestnut-brown and olive-green; upper tail-coverts olive, tinged with yellow and marked with dusky-grey streaks; tail-quills brown, outside feathers edged with white; chin and throat black; upper part of breast dull olive, crossed below by a chestnut band; belly and under tail-coverts dusky-yellow; sides dull olive streaked with dark brown.

The female is slightly smaller; she lacks the bright yellow stripes on the sides of the head and throat; the black upon the chin is replaced by yellowish-brown; crown dull olive streaked with black; back and upper-parts reddish-brown streaked with black; under-parts dirty yellow, also streaked with black. (See how to distinguish this species from Yellow-hammer, p. 518.)

*Situation and Locality.*—In brambles, furze bushes, and sometimes quite on the ground, in similar situations to the Yellow Bunting. On commons and cultivated lands well studded with trees and hedgerows in the south and west of England; also in Wales, local.

*Materials.*—Dry grass, roots, moss, and leaves, with generally an inner lining of hair.

*Eggs.*—Four or five. Dull bluish or cinerous white, spotted, blotched, streaked, and veined irregularly with very dark brown, and underlying markings of grey. Size about .86 by .65 in. (*See* Plate II.)

*Time.*—May, June, and July.

*Remarks.*—Resident. Notes: call, *zi-zi-za-zirr;* song, *zis-zis-zis-gor-gor-gor,* according to Bechstein, whilst another authority describes it as *tutt, tutt, tutt, tutt, tutt, tutt.* Local and other names: Black-throated Yellow-hammer, French Yellow-hammer. Sits closely.

---

## BUNTING, COMMON. *See* BUNTING, CORN.

---

## BUNTING, CORN. *Also* COMMON BUNTING.

(*Emberiza miliaria.*)

Order PASSERES; Family EMBERIZIDÆ (BUNTINGS).

*Description of Parent Birds.*—Length a little over seven inches. Bill short, conical, strong, and pale yellow-brown, with a stripe of dark brown on the top of the upper mandible. Irides dark hazel. Head, neck, back, rump, and upper tail-coverts light brown, inclining to olive, each feather being streaked in the centre with dark brown. Wings dark brown, the feathers being edged with a lighter tinge of the same colour. Tail slightly forked and dark brown with light edges. Chin, throat, breast, belly, vent, and under tail-coverts dull whitish-brown; the sides of the neck and the breast are marked with triangular spots of dark brown; sides streaked with the same colour. Legs, toes, and claws pale brown. The bird is thick and bulky in appearance and of sluggish habits.

Female similar to male.

*Situation and Locality.*—On or near the ground, amongst coarse grass, on a bank, among the grass, under a hedge, in a low bush, amongst growing

CORN BUNTING'S NEST AND EGGS.

corn, brambles, clover, and peas; in grass-fields, pastures, clover-fields, and similar places locally distributed throughout the United Kingdom.

*Materials.*—Straw and coarse hay or grass-stems outside, lined with fibrous roots, fine grass, and sometimes horsehair.

*Eggs.*—Four to six. The ground-colour varies from dull purplish-white to pale buff, blotched, spotted, and streaked with light to dark purplish-brown, and underlying markings of grey. They are variable in size, but run larger than those of any other Bunting breeding with us. Size about .96 by .71 in. (*See* Plate II.)

*Time.*—May and June. Sometimes as early as end of April and as late as July.

*Remarks.*—Resident, but numbers swollen during winter months by Continental arrivals. Notes: *chuck* or *chit.* Local and other names: Common Bunting, Bunting Lark, Ebb. Sits fairly close.

## BUNTING, REED. *Also* REED SPARROW.

(*Emberiza schœniclus.*)

Order PASSERES; Family EMBERIZIDÆ (BUNTINGS).

FEMALE REED BUNTING.

*Description of Parent Birds.*—Length about six inches. Bill short, conical, and dusky-brown on the upper mandible, lighter on the lower. Irides hazel. Head velvety black, bounded by a white collar, which commences near the gape and, descending the sides of the neck as far as the breast, passes round the back thereof. Back and wings rich brownish-black, the feathers being margined broadly with reddish-brown and tawny-grey; wing-quills dusky, narrowly bordered with tawny-red. Rump and upper tail-coverts black, tinged with rusty-grey. Tail-quills brownish-black, bordered on the outsides with white, and slightly forked. Chin and throat black, broad in the centre, and pointed on the lower part of the breast. Breast on either side of the black portion white, also belly and under-parts; sides and flanks tinged and streaked with brown. Legs, toes, and claws dusky-brown.

The female is smaller, and differs considerably in her plumage. Her head is brown instead of black; the white collar of the male is dusky-brown in her case. Chin, throat, and breast dull white.

*Situation and Locality.*—Generally near the ground amongst long grass, rushes, nettles, and sedges. I have once or twice met with it in low thorn bushes, amongst grass and weeds growing about the stunted branches which had been cropped by sheep. Our illustration was procured in July amongst the sprouts and long grass growing round the stump of a felled tree on the banks of the Mole, in Surrey. The nest is generally found close to sluggish streams, ponds, swamps, and bodies of water, though I have frequently met with it at considerable distances from water. It is said to have been found in trees at a height of eight or nine feet. The nest may always be known from that of the Reed Warbler by the fact that it is never suspended. The bird breeds in nearly all suitable localities throughout the British Isles. I have met with it nesting in the heather by loch-sides in the Outer Hebrides and in the north of Caithness.

REED BUNTING'S NEST AND EGGS.

*Materials.*—Dried grass and moss, with a lining of finer grass, hair, and the feathery tops of reeds.

*Eggs.*—Four to seven, generally five. Purplish-grey or pale olive to pale purple-brown in ground-colour, spotted and streaked with rich dark purple-brown, generally distributed over the egg. It is the smallest of the Buntings' eggs found in this country, and the veins are shorter and thicker than those of the Yellow Bunting. Size about .77 by .57 in. (*See* Plate II.)

*Time.*—March, April, May, June, and July.

*Remarks.*—Resident, and partially migratory. Notes: song, *te, te, tu, te,* diversified by an occasional discordant *ruytsh;* alarm note, a sharp twitter. Local and other names: Reed Sparrow, Passerine Bunting, Black Bonnet, Chink, Water Sparrow, Black-headed Bunting (a name which, properly speaking, belongs to a different species altogether), Mountain Sparrow. Sits closely. The male bird does his share in the work of incubation and feeding of the chicks.

MALE REED BUNTING FEEDING YOUNG.

## BUNTING, SNOW.

(*Plectrophanes nivalis.*)

Order PASSERES; Family EMBERIZIDÆ (BUNTINGS).

*Description of Parent Birds.*—Length about seven inches; bill short, conical, and black. Irides hazel. Head and neck white (in some specimens the crown and nape are mottled with black); back velvety black; rump and upper tail-coverts white, some of the feathers being slightly bordered with brownish-white; wings black on the shoulder or point, white through the middle, and black on outer half and tips; tail slightly forked, white on the outside and black in the middle; chin, throat, breast, belly, vent, and under tail-coverts pure white; legs, toes, and claws black.

The female is rather smaller, has the white on the head and neck more mottled with black, and her colours generally are not so pure. Very few specimens of the bird have been secured in this country in its breeding plumage.

*Situation and Locality.*—In crevices and chinks of rock, or amongst loose stones. The bird sometimes breeds on the high hills and mountains of the north of Scotland, in the Orkneys and Shetlands; but very few nests indeed have ever been found. I have sought for it in a good many localities without success, and need hardly say that I shall feel greatly obliged to any ornithologist who will at any time give me an opportunity of photographing a nest and eggs *in situ.*

*Materials.*—Dead grass and roots, with an inner lining of finer grass, hair, wool, and feathers where procurable. The same nest is said to be used more than once.

*Eggs.*—Four to eight, more often four to six; ground-colour greenish- or bluish-white, sometimes greyish-white, pale bluish-grey, or pale greenish-blue, spotted, splashed, and streaked with deep brownish-red, and a few spots and streaks of a darker tint on the top of these; occasionally underlying markings of pale grey or yellowish-brown. Size about .86 by .64 in. (*See* Plate II.)

*Time.*—May, June, and July.

*Remarks.*—A winter visitor, a few pairs resident. Notes: sweet and tinkling. Local names: Snow Flake, Snow Fleck, Snow Fowl, Tawny Bunting, Greater Brambling, Lesser Mountain Finch, Great Pied Mountain Finch, Brambling (a name belonging to another bird altogether). A close sitter.

---

## BUNTING, YELLOW. *See* YELLOW HAMMER.

---

## BUZZARD, COMMON.

(*Buteo vulgaris.*)

Order ACCIPITRES; Family FALCONIDÆ (FALCONS).

*Description of Parent Birds.*—Length about twenty-two inches. Beak short, much curved, strong, and blue-black in colour. Bare skin round the base of the beak yellow. Irides yellow. Crown, nape, back wing-coverts, and upper side of tail-quills clove or ferruginous brown, with large longitudinal spots and dashes; the tail being barred with black and ash-colour, and at the end dusky-white. Wing-quills brownish-black. Chin and throat yellowish- or dusky-white. Breast, belly, and thighs greyish- or yellowish-white, streaked and spotted with yellowish-brown. Under tail-coverts white. Legs and toes yellow; claws black.

The female is darker than the male, and often larger. The colour of plumage in both sexes is subject to great variation.

COMMON BUZZARDS' BREEDING HAUNT.

*Situation and Locality.*—In the forked branches of a tree, sometimes on a horizontal branch at a little distance from the trunk. Also in high, inaccessible maritime cliffs and tall crags in wild secluded districts of England, Wales, Scotland, and Ireland. The bird will often adopt an old crow's nest, and generally returns to the same breeding-place year after year. It is still fairly common in some parts of Wales, where I have seen as many as three pairs on the wing at once.

*Materials.*—Sticks and twigs in liberal quantities, lined with hay, wool, and leaves, sometimes scraps of down.

*Eggs.*—Two to four, generally three. Sometimes dingy white and unspotted, and in other instances greenish- or bluish-white, spotted, blotched, and streaked with red-brown and pale rust-colour.

Very variable in regard to size and coloration. Average about 2.16 by 1.72 in. (*See* Plate VII.)

*Time.*—April and May.

COMMON BUZZARD'S NEST AND EGGS.

*Remarks.*—Resident. Note: a monotonous and plaintive *pe-e-i-o-oo.* Local and other names: Buzzard, Puttock. Not a very close sitter, except when incubation is advanced.

## BUZZARD, HONEY.

(*Pernis apivorus.*)

Order Accipitres ; Family Falconidæ (FALCONS).

Though never a numerous species in our islands, this bird did at one time breed in several different parts of the country. It is probable that the high price set upon its eggs, and the senseless persecution of the harmless parent birds, have almost, if not quite, banished the Honey Buzzard from its last stronghold—viz. the New Forest.

---

## BUZZARD, ROUGH-LEGGED.

(*Archibuteo lagopus.*)

Order Accipitres ; Family Falconidæ (FALCONS).

This bird is said to have bred in both England and Scotland, but as the instances are rare, and the information concerning them is scant, it can claim but little attention in a work of this character.

---

## BURROW DUCK. *See* Sheldrake.

---

## CAPERCAILLIE.

(*Tetrao urogallus.*)

Order Gallinæ ; Family Tetraonidæ (GROUSE).

*Description of Parent Birds.*—Length varies between thirty-three and forty inches. Bill short, much curved, strong, and of a pale horn colour. Irides hazel. Over each eye is a piece of naked red skin. Head, neck, back, rump, and upper

CAPERCAILLIE'S NEST AND EGGS.

tail-coverts brownish-black in ground-colour, finely freckled with ash-grey spots. Wings dark chest-nut-brown, minutely speckled with dusky spots, except quills, which are dusky. Where the shoulder or point of the wing meets the body is a little patch of white; the scapulars are also tipped with the same colour. Tail rounded at the tip, the feathers being dusky, spotted sparingly with light grey on the sides; a few shorter ones lying over the principal quills are tipped with white. Chin and throat dull black, the feathers being some-what elongated. Upper breast dark glossy green, lower breast and all under-parts black, spotted sparingly with white about the thighs and vent. Legs covered with brown hair-like feathers; toes and claws black.

The female measures about twenty-six inches in length, and differs very considerably in the colour of her plumage. The feathers of her head, neck, back, wings (except quills, which are dusky), rump, upper tail-coverts, and tail tawny-brown, barred with blackish-brown and tipped with white. Throat tawny-red; breast of a lighter tinge, spotted spar-ingly with white; belly and under-parts generally barred with pale tawny and black, the feathers being tipped with greyish-white. Legs greyish-brown; toes and claws pale brown.

*Situation and Locality.*—On the ground, gener-ally at the foot of a tree, as shown in our illus-tration, sometimes under a bush or bramble, amongst long grass or heather, in Scotch fir, larch, and spruce forests; also, but more sparsely, in oak and birch forests, through the midlands of Scotland.

*Materials.*—Small quantities of dead grass,

leaves or pine needles used as a lining to the hollow, scraped or chosen, in the ground.

*Eggs.*—Six or eight to twelve or fifteen. Pale reddish-yellow, spotted all over with two shades of darker orange-brown. Size about 2.2 by 1.6 in. (*See* Plate XV.)

*Time.*—April, May, and June.

*Remarks.*—Resident. This bird became extinct in Britain towards the end of the eighteenth century, and was re-introduced from Sweden in 1837; since that time it has thriven and spread in Scotland. Call of male: *peller, peller, peller.* The note of the female is a hoarse *gock, gock, gock.* Local and other names: Wood Grouse, Ceiliog Coed (of the ancient British), Cock of the Woods, Great Grouse, Cock of the Mountain, Caper, Capercally, Capercailzie, Capercali. Sits closely.

---

## CHAFFINCH.

(*Fringilla cœlebs.*)

Order Passeres; Family Fringillidæ (FINCHES).

*Description of Parent Birds.*—Length about six inches. Bill shortish, strong, conical, pointed, and dark blue. Irides hazel. Forehead black; crown, hinder part of head, and a part of the sides of the neck bluish-ash. Back reddish-brown; rump and upper tail-coverts greenish. Lesser wing-coverts white; greater black tipped with white, thus forming two conspicuous bars across the wings; quills dusky, bordered with greenish-yellow on the

MALE CHAFFINCH.

outer webs and marked with greyish-white on both webs near the base. Tail-quills black, tinged with grey on the two middle feathers, and the two outer ones on each side marked with white. Chin, cheeks, throat, breast, belly, and under-parts reddish, chest nut-brown, paler on the belly, vent, and under tail-coverts. Legs, toes, and claws dusky.

The female is smaller, and her head, neck, and upper part of the back greyish-brown; the rump and upper tail-coverts are not so bright, and her under-parts are brownish-white, tinged with red upon the breast. The white bars upon her wings are not so conspicuous.

*Situation and Locality.*—In the forks of small trees, on branches and twigs of whitethorns, fruit-trees, in hedges, gorse-bushes, and other kinds of trees in orchards, spinneys, on commons, and almost anywhere and everywhere where there are trees or shrubs throughout the United Kingdom.

*Materials.*—Moss, wool, lichens, and cobwebs, beautifully felted together, and lined with hair, feathers, and down. The nest is cup-shaped, deep, and wonderfully made in every respect. It is compact, neat, well woven together, and securely fastened to the situation chosen. The bird shows a great deal of sagacity in its outside adornment. I have seen specimens in lichen-covered or grey-barked trees that were smothered with bits of lichen and spiders' nests, and have known bits of old newspaper used for the purpose. On the other hand, I have met with many whereon none of these materials appeared, because their surroundings did not call for them to produce any harmonising effect.

CHAFFINCH ON NEST.

*Eggs.*—Four to six, generally five. Pale greenish-blue, generally suffused with faint reddish-brown and spotted and streaked with dirty reddish-brown of various shades. Occasionally specimens are met with suffused with purplish-buff all over, or of a uniform pale greenish-blue without markings of any kind. Size about .75 by .58 in. (*See* Plate II.)

*Time.*—April, May, June, and July.

*Remarks.*—Resident, and partially migratory. I have noticed that the cocks, almost without exception, leave the Yorkshire dales in winter. Notes: *spink-spink, yack-yack, treef-treef.* Its song is a joyous, ringing trill. Local and other names: Bullspink, Scobby, Skelly, Spink, Twink, Pink, Shellapple, Shelly, Shilfer, Wet Bird, Buckfinch, Beechfinch, Whitefinch, Copperfinch, Horsefinch. A very close sitter. I have seen the bird's tail pulled out by a lad attempting to catch the hen on her nest; yet, quite undaunted, she returned, hatched out her five eggs, and reared her young. The male bird helps the female to feed the chicks both in the nest and after they are fledged.

YOUNG CHAFFINCH.

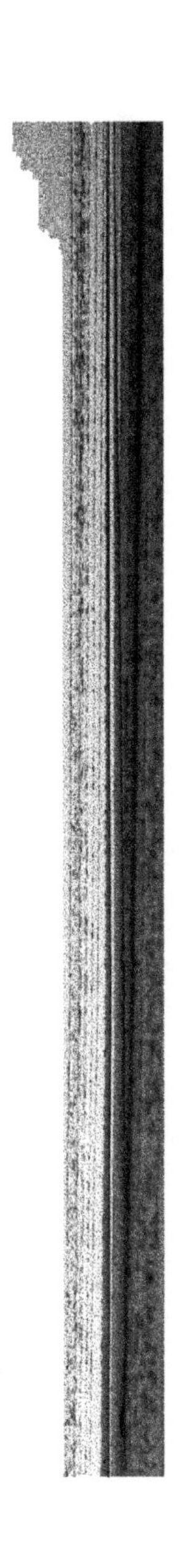

(*See p.* 332.) (*See p.* 52 )

STARLING.
(*See p.* 386.)

RAVEN.
(*See p.* 314.)

MAGPIE.
(*See p.* 216.)

HOODED CROW.
(*See p* 55.)

NOTE.—*In referring to the eggs the above nane*

## CHIFFCHAFF.

(*Phylloscopus rufus.*)

Order PASSERES; Family SYLVIIDÆ (WARBLERS).

*Description of Parent Birds.*—Length about four and three-quarter inches. Bill moderately long, nearly straight, and dark brown in colour. Irides brown. Over the eye is a pale yellowish-brown streak, which becomes much lighter behind the ear-coverts. Crown, neck, back, and upper tail-coverts dull olive-green tinged with yellow. Wing-quills dark greyish-brown edged with olive-green; tail-feathers somewhat similar. Chin, throat, breast, and under-parts dull yellowish-white. Legs, toes, and claws blackish-brown.

The female is very similar in all respects.

*Situation and Locality.*—On or near the ground, in woods, on hedge-banks amongst tall, rank grass, supported by brambles and slender bushes. The bird breeds pretty generally throughout the south and middle of England, but less frequently in the northern counties and Scotland. It is numerous in many parts of Wales and is met with in Ireland.

*Materials.*—Dead grass, withered leaves, and moss, with an inner lining of hair and a liberal quantity of feathers. The nest is oval, or nearly so, domed, and has an entrance hole at the side and near the top.

*Eggs.*—Five to seven, white, somewhat sparingly spotted with dark purplish-brown. The spots vary in size, intensity, and numbers, but, as a rule, they are darker, fewer, and larger than those found upon the eggs of the Willow Wren. Size about .6 by .48 in. (*See* Plate III.)

CHIFFCHAFFS AT NEST.

*Time.*—April, May, and June.

*Remarks.*—Migratory, arriving in March and departing in October. Notes: *chiff-chaff*, which has also been represented as *chip-chop*, *chivy-chavy*, *choice and cheap*. Alarm note, *whoo-id* or *whoo-it*. Local and other names: Least Willow Wren, Lesser Pettychaps, Choice and Cheap. A close sitter.

CHIFFCHAFF'S NEST AND EGGS.

---

## CHOUGH.

(*Pyrrhocorax graculus.*)

Order PASSERES; Family CORVIDÆ (CROWS).

*Description of Parent Birds.*—Length about sixteen inches. Bill rather long, slightly curved, and orange-red. Irides hazel. The whole of the plumage is black, glossed with purple; legs and toes red; claws strongly hooked, and black.

The female is not as large as the male, and her bill is shorter.

*Situation and Locality.*—In clefts and fissures of sea cliffs, holes in old ruins, and in caves. The bird is said to still breed very sparingly in Cornwall, Devonshire and Lundy Island. It also nests, though much persecuted, along the Welsh coast, in the Isle of Man, and more plentifully on the west coast of Scotland, and in some parts of Ireland; but everywhere seems to be on the decrease—a distressing fact which it is to be earnestly hoped all true lovers of British birds will do their very utmost to assist in checking, as there is every reason to believe that this interesting bird is much rarer than it is generally supposed to be.

*Materials.*—Sticks, dead heather stalks, dry grass, wool, and occasionally hair.

*Eggs.*—Four or five, occasionally six. Dirty or creamy white, sometimes faintly tinged with green or blue, spotted with light brown and ash-grey. Markings variable, both in regard to size and distribution. Size about 1.52 by 1.1 in. (*See* Plate I.)

*Time.*—May.

*Remarks.*—Resident. Notes: *creea, creea,* rendered by Mr. Seebohm as *khee-o, khee-o.* It utters a quick, chattering noise at times, like a Starling. Local and other names: Cornish Daw, Cornish Chough, Cornwall Kae, Market Jew Crow, Chauk Daw, Red-legged Crow, Killigrew, Hermit Crow, Cliff Daw, Gesner's Wood Crow. Gregarious. It is thought by some ornithologists that the Jackdaw has driven the Chough away from some of its old haunts. Not a very close sitter, and noisy when intruded upon.

INTERIOR OF CAVE CONTAINING CHOUGH'S NEST

## COOT, COMMON.

(*Fulica atra.*)

Order FULICARIÆ; Family RALLIDÆ (RAILS).

COOT ON NEST.

*Description of Parent Birds.*—Length about eighteen inches. Bill of medium length, nearly straight, and dull white tinged with red. There is a smooth, naked white patch on the forehead which readily distinguishes this bird from the Waterhen, with its red shield. Irides hazel. Under the eye is a narrow, curved line of white. The whole of the plumage is black with exception of white on the bend of the wing and narrow bar formed by the white tips of the secondaries. The under-parts are tinged with bluish-grey. Legs, toes, and scallop-shaped lobes on either side of the latter, dark green. Above the knee is an orange-coloured garter.

The female is similar in size and appearance to the male.

*Situation and Locality.*—Amongst reeds, rushes, osiers, and other aquatic herbage in marshes, by the sides of ponds, reservoirs, large sluggish rivers, and lakes. They are generally built up from the bottom, but are sometimes simply moored to surrounding objects, and, becoming detached by wind or floods, float about without any apparent

COOT'S NEST AND EGGS.

inconvenience to the builder. I recollect once seeing an instance of this kind on a large reservoir in South Yorkshire. The bird breeds in suitable places in nearly all parts of the British Isles.

*Materials.*—Decaying sedges, flags, reeds, and rushes, and although not very elegant is wonderfully strong. It is a very pretty sight to see this bird pulling up decayed weeds and swimming with them to its half-constructed nest.

*Eggs.*—Seven to ten. As many as fourteen or fifteen have been found. Dingy stone-colour or dull buff, spotted and speckled all over with nutmeg brown. Size about 2.1 by 1.5 in. (*See* Plate XIV.)

*Time.*—April, May, June, and July.

*Remarks.*—Resident, and partially migratory. Note: a clear ringing *ko*. Local and other names: Common Coot, Bald Coot. Sits lightly.

COOT.

## CORMORANT.

(*Phalacrocorax carbo.*)

Order Steganopodes ; Family Pelecanidæ (PELICANS).

ADULT CORMORANT AND YOUNG.

*Description of Parent Birds.*— Length about thirty-six inches. Bill rather long, straight except at the tip, where it is hooked, and pale brown. Irides green. Crown, nape, and a portion of the neck black, intermixed with a number of very narrow white feathers almost like hairs. The feathers at the back of the head are elongated into a kind of crest. The feathers on the back and of the wing-coverts are dark brown bordered with black. Wing and tail quills black. Round the base of the bill and chin bare and yellow, bordered with white on the latter. Neck all round below the portion streaked with white, breast, belly, and under-parts rich bluish-black, except on either thigh, where there is a patch of white. Legs, toes, membranes, and claws black. The female is said not to be so large as the male, but has her crest often longer.

*Situation and Locality.*—On ledges of cliff by the sea, and also inland, on low rocky islands and reefs, sometimes in trees. The bird breeds pretty generally round our coasts wherever suitable cliffs and rocks are to be found, and still maintains inland colonies at a few places.

CORMORANTS' NESTS AND EGGS.

*Materials.*—Sticks, twigs, and coarse grass or seaweed, depending upon locality. It is a large, high nest.

*Eggs.*—Three generally, but sometimes as many as five or six. The real shell is a pale blue, but this is usually hidden by a thick coating of chalk which can easily be scraped off. Sometimes the real shell shows through the casing of lime. Size about 2.6 by 1.62 in.

*Time.*—April, May, June, July, and sometimes even in August.

*Remarks.*—Resident. Note: the bird is remarkably silent, but occasionally utters "a harsh croak" according to Seebohm. Local and other names: Crested Cormorant, Corvorant, Great Black Cormorant, Cole Goose, Sea Crow, Scart Brongie, Norie, Isle of Wight Parson. Sits closely or lightly according to situation.

CORMORANTS ON NESTS

## CRAKE, BAILLON'S.

(*Porzana bailloni.*)

Order FULICARIÆ ; Family RALLIDÆ (RAILS.)

A rare bird, whose nest and eggs have only been found about twice in the British Isles, and in neither case within recent times.

---

## CRAKE, CORN.

(*Crex pratensis.*)

Order FULICARIÆ ; Family RALLIDÆ (RAILS).

CORN CRAKE ON NEST.

*Description of Parent Birds.*—Length about ten and a half inches. Bill of medium length, nearly straight, thick at the base, and light brown. Irides hazel. Over the eyes and ear-coverts, also cheeks, ash-grey. Head, neck, back part of wings, tail-coverts, and feathers reddish-brown, with a long dusky streak in the centre of each. Wing-coverts rich bay, quills reddish-brown. Breast, belly, and under-parts pale yellowish-brown or light buff, barred on the sides with two shades of reddish-brown. Legs, toes, and claws are light yellowish-brown.

The female is rather inferior in size to the

CORN CRAKE'S NEST AND EGGS.

male, and the colours on her cheeks and wings are less distinct.

*Situation and Locality.*—On the ground amongst mowing grass, clover, willow beds, and standing corn in suitable parts of the United Kingdom. I have never met so many birds of this species anywhere as in the Outer Hebrides, where I have found as many as four nests in a single day.

*Materials.*—Strong stems of dead grass and leaves, with an inner lining of finer grass, much of which is added by the bird from day to day whilst she is laying.

*Eggs.*—Seven to ten as a rule. I have on a good many occasions found eleven, and as many as twelve and even fifteen have been met with, however. Pale reddish-white or light buff, spotted, freckled, and blotched with red-brown of various shades and ash-grey. Size about 1.4 by 1.1 in. (*See* Plate XIV.)

*Time.*—May and June, but nests containing eggs have been reported even as late in the year as September.

*Remarks.*—Migratory, although individuals remain all the year round in Ireland. Arrives in April and May and departs in September and October, but stragglers remain later. Notes: *crake, crake.* Local and other names: Landrail, Meadow Crake, Corn Creak, Draker Hen. The bird sits closely, and, as a consequence, individuals sometimes get their heads cut off by the mower's scythe or machine. It is at the time of revising this work (1908) a decreasing species in some parts of the British Islands.

## CRAKE, SPOTTED.

(*Porzana maruetta.*)

Order FULICARIÆ ; Family RALLIDÆ (RAILS).

*Description of Parent Birds.*—Length about nine inches. Bill of medium length, straight, thick at the base and pointed at the tip, yellowish-brown in colour, with a brighter and redder tinge towards the base. Irides brownish-hazel. Crown hazel-brown spotted with black in the middle ; over each eye is a patch of dull blue-grey ; sides of head, nape and sides of neck olive-brown spotted with white. The feathers of the back are black, broadly edged with dark olive-brown, and streaked up and down with fine lines of white. Wing-coverts olive-brown ; quills dark brown, mottled and barred with white ; rump, upper tail-coverts, and tail-quills black, bordered with dark reddish-brown and spotted with white. Chin grey-brown ; throat and breast dusky-brown mottled with white ; sides and flanks greyish-brown barred with white ; under tail-coverts buffish-white, legs and toes yellowish-green (toes long), claws brown.

The female is a trifle smaller, and not quite so distinctive in coloration.

*Situation and Locality.*—In a tussock of sedge, amongst reeds and other vegetation growing in marshes, bogs, and wet, swampy ground. The foundation is generally in water. Sparingly along the east coast counties, in Wales, Cumberland, one or two suitable parts of Scotland, and Ireland.

*Materials.*—Coarse aquatic plants, such as reeds and flags in somewhat liberal quantities, and lined with dry grass.

*Eggs.*—Seven to twelve. White, yellowish-grey, or ocherous. Some authorities describe them as being occasionally white tinged with green, or grey tinged with pink, spotted with dark reddish-brown, and underlying markings of grey. Size about 1.3 by .9 in. (*See* Plate XIV.)

*Time.*—May is the principal laying month; however, eggs have been found in April and right through June.

*Remarks.*—Migratory and resident. The first kind of birds arrive in March and depart in October. Notes: *whuit, whuit.* Local and other names: Spotted Rail, Water Rail, Water Crake (this name is also applied to the Dipper), Spotted Gallinule, Spotted Water Hen. Slips quietly off nest and hides amongst surrounding vegetation.

---

## CREEPER, TREE.

(*Certhia familiaris.*)

Order PASSERES; Family CERTHIDÆ (CREEPERS).

YOUNG TREE CREEPER.

*Description of Parent Birds.*—Length about five inches. Bill rather long, curved downwards, dark brown on the top, and dirty white, tinged with yellow, underneath. Irides hazel. Crown dark brown, spotted and streaked with pale brown. Back of neck, back, and rump tawny-brown, mixed with ash-grey. Wing-coverts brown, tipped with greyish-yellow; quills variegated with brown and black, some of them tipped with light grey. Tail tawny-brown, the feathers being strong in the shaft,

EXAMINING SITE OF TREE-CREEPER'S NEST.

and, from their help to the bird in climbing and holding on to the bark of trees, often worn quite bare at the ends. Chin, throat, breast, and belly greyish-white, inclining to rusty reddish-white on the flanks and vent. Legs, toes, and claws, which are very long, light brown.

The female is similar in size and colour to the male.

*Situation and Locality.*—In a hole in a tree; behind a loose piece of bark still clinging to a decayed tree; amongst piles of stacked timber; in niches and crevices of buildings, and behind half-detached pieces of plaster. The one in our illustration was placed behind a sound piece of the outer shell of a decayed pollard. The bird could either slip off from the front, or up a kind of chimney, having its exit just under the face of the inquisitive onlooker. Found in nearly all well-wooded districts throughout the British Isles.

*Materials.*—Fine twigs, dead grass, sometimes little chips of decayed wood, wool, moss, feathers, and rabbits' down.

*Eggs.*—Six to nine, white, spotted and speckled with reddish-brown and sometimes dullish purple spots, generally in a kind of zone round the larger end, but occasionally more distributed. Size about .66 by .47 in. (*See* Plate III.)

*Time.*—April, May, and June.

*Remarks.*—Resident. Notes: song not often heard, but high, shrill, and not unpleasant. Local and other names: Creeper, Tree Climber, Common Creeper. Sits very closely.

| | | | |
|---|---|---|---|
| (*See p.* 11.) | (*See p.* 14.) | (*See p.* 50.) | (*See p.* 19.) |
| ELLOW HAMMER. (*See p.* 520.) | TREE SPARROW. (*See p.* 380.) | REED BUNTING. (*See p.* 17.) | HAWFINCH. (*See p.* 184.) |
| LINNET. (*See p.* 213.) | HOUSE SPARROW (*See p.* 372.) | GREENFINCH. (*See p.* 145) | BULLFINCH. (*See p.* 8.) |
| TWITE. (*See p.* 443.) | CHAFFINCH. (*See p.* 28.) | LESSER REDPOLE (*See p.* 318.) | GOLDFINCH. (*See p.* 130.) |
| MEADOW PIPIT. (*See p.* 282.) | WOODLARK. (*See p.* 501.) | SKYLARK. (*See p.* 367.) | ROCK PIPIT. (*See p.* 286.) |

NOTE.—*In referring to the eggs the above names should be read from left to right.*

## CROSSBILL.

(*Loxia curvirostra.*)

Order PASSERES; Family FRINGILLIDÆ (FINCHES).

*Description of Parent Birds.*—Length from six and a half to seven inches. Beak rather large, upper mandible turned down and lower one up. They do not lie in consequence in a straight line over each other, but cross like the blades of a pair of scissors. The Crossbill varies more in coloration according to age, sex, and individual than perhaps any other British bird.

Swaysland gives the following description: "When young the male birds are greenish-brown, with a tinge of olive, the whole being speckled with darker brown; they are, however, lighter upon the under-parts; but after the first moult a red tinge prevails, occasioned by the tipping of the feathers with that hue. The red is much darker upon the upper-parts. At the second moulting these colours are lost, and the bird's plumage becomes an olive-brown, shaded over with greenish-yellow upon the back, though it is much lighter upon the under-parts, and is speckled with orange upon the breast and rump.

"The females are, however, either grey with a little green on the head, breast, and rump, or else speckled in an irregular manner with those colours."

Professor Newton, in describing the male with the second dress on, says: "A red male that had completed his first autumnal moult had the bill dull reddish-brown, darkest towards the tip of the upper mandible; irides dark brown; the head, rump, throat, breast, and belly tile-red; the feathers on the back mixed with brown, producing a chestnut brown; wing and tail feathers nearly uniform dark brown; vent and lower tail-coverts greyish-white; legs, toes, and claws dark brown. . . . Young females, after their first striated dress, acquire a greenish-yellow tint on the crown and the lower parts of the body mixed with greyish-brown; the rump and upper tail-coverts of primrose yellow tinged with green; wings, tail, and legs as in the male."

A celebrated Continental authority, writing upon the matter, says: "If the Crossbills are grey or speckled, they are young; if red, they are one year old and have just moulted; if carmine, they are just about to moult for the second time; if spotted with red and yellow, they are two years old and in full feather."

*Situation and Locality.*—On the branches of Scotch and other fir-trees, sometimes quite close to the bole or stem, at others some distance away on a horizontal branch at heights varying from twenty to forty feet from the ground. Generally in plantations of cone-bearing trees over the greater portion of England, Scotland, and in Ireland, where suitable plantations are to be met with. The bird is very sparsely distributed, and uncertain in its patronisation of recognised breeding haunts. It is most numerous in the central counties of Scotland.

*Materials.*—Twigs, roots, coarse dead grass, lined internally with finer grass, hairs, and feathers. It is similar in construction and appearance to that of the Greenfinch.

*Eggs.*—Four or five. White, sometimes faintly tinged with pale blue, very sparingly speckled with

CROSSBILL'S NEST AND EGGS.

reddish-brown and pale brown. Average size about .9 by .67 in. (*See* Plate II.)

*Time.*—February, March, and April; sometimes as late even as July. Continental authorities say from December to April.

*Remarks.*—A winter visitor, but a few pairs stay to breed. Notes: call, *chip-chip-chip*, or *jip-jip*. Other notes used whilst flying from tree to tree, *soc-soc-soc*. Local and other names: Common Crossbill, Shell Apple, European Crossbill. A very close sitter.

---

## CROW, CARRION.

(*Corvus corone.*)

Order PASSERES; Family CORVIDÆ (CROWS).

CARRION CROW'S NEST AND EGGS.

*Description of Parent Birds.* — Length about eighteen inches. Bill fairly long, strong, and black. The base of the beak is covered with bristles, which stand forward. These bristles and its hoarser note distinguish it from the rook. Irides dusky. The whole of the plumage is black, glossed above with a lustrous greenish sheen. Legs, toes, and claws black. The female is about the same size, but lacks a little of the metallic lustre which characterises the male.

*Situation and Locality.*—In high trees, generally on a large branch near the bole, and at a good height from the ground, on the outskirts of woods

CARRION CROW'S NEST.

and plantations; sometimes on ledges of cliffs. I have on more than one occasion found a nest in a thorn bush not more than twelve feet in height. In England, Wales, Scotland, and the north of Ireland. The bird is nowhere very numerous, as

YOUNG CARRION CROWS.

its predatory habits make for it an uncompromising enemy in the gamekeeper.

*Materials.*—Sticks and mud, lined with grass, wool, horse and cowhair.

*Eggs.*—Four or five, occasionally as many as six, grey-green, blotched and spotted with ash-colour or smoky-brown; sometimes they are found quite blue, and minus spots of any kind. They are similar to those of the Rook and Raven, but larger than the former and smaller than the latter, and the position of the nest generally suffices to distinguish them. Size about 1.65 by 1.2 in. (*See* Plate I.)

*Time.*—April, May, and June.

*Remarks.* — Resident. Numbers increased in

winter by arrivals from Continent. Note: a hoarse croak. Local and other names: Crow Mussel (from its habit of eating mussels), *D*oup, Gor Crow, Minden Crow, Black-nebbed Crow. Sits lightly, and generally in such a position as to command a good surrounding view.

---

## CROW, GREY. *See* CROW, HOODED.

---

## CROW, HOODED. *Also* GREY CROW *and* ROYSTON CROW.

(*Corvus cornix.*)

Order PASSERES; Family CORVIDÆ (CROWS).

HOODED CROW.

*Description of Parent Birds.*—Length about twenty inches. Beak moderately long, pointed, strong, and black; the base is covered by stiff projecting feathers. Irides dusky. Head, throat, wings, and tail a shining blue-black. Nape, back, rump, and underparts generally dark slaty-grey. Legs, toes, and claws a shining black.

The female is a little smaller in size, and the slaty-grey parts of her plumage are tinged with brown.

*Situation and Locality.*—Rocks, cliffs, and trees. An instance has been recorded of this bird building on the roof of a crofter's hut. I have on several occasions found nests in deep

HOODED CROW'S NEST IN HEATHER ON THE GROUND

heather on the ground, as depicted in our full-page illustration. In Ireland as far south as Killarney, the Isle of Man, the mainland of Scotland, and the islands to the west and north.

*Materials.*—Sticks, twigs, heather, and ling, with an inner lining of roots, moss, wool, hair, or feathers. Sometimes they are composed entirely of dead seaweed and wool.

*Eggs.*—Three to six; generally five, grey-green in ground-colour, blotched and spotted with varying shades of olive- or greenish-brown. Variable both in regard to size, shape, ground-colour, and colour of markings. Size about 1.65 by 1.2 in. (*See* Plate I.)

*Time.*—March, April, May, and June.

*Remarks.*—Resident, but subject to southern movement in winter, when numbers are greatly increased by arrivals from the Continent. Note: a hoarse croak. Local and other names: Hoody, Dun Crow, Grey Crow, Bunting Crow, Royston Crow, Greyback, Norway Crow, Kentish Crow, Scarecrow. The bird is a light sitter, and sometimes interbreeds with the Carrion Crow.

YOUNG HOODIES.

## CROW, ROYSTON. *See* CROW, HOODED.

*See* CROW, HOODED.

---

## CUCKOO.

(*Cuculus canorus.*)

Order PICARIÆ; Family CUCULIDÆ (CUCKOOS).

ADULT CUCKOO.

*Description of Parent Birds.*—Length about fourteen inches. Bill rather short, slightly curved downwards, and black, turning yellowish at the base. Irides yellow. Head, nape, back, and upper parts generally dark ash-colour. Wing-quills dusky, barred with white for some distance on their inner webs. Tail-quills greyish-black, especially the middle feathers, tipped with white, and marked with white spots. Chin, throat, and upper breast pale ash-grey; lower breast, belly, vent, and under tail-coverts white, marked with wavy, transverse bars of black; the two last parts often have a reddish-brown tinge. Legs, toes, and claws yellow.

The female is very similar in appearance to the male, but a little smaller in size.

*Situation and Locality.*—Deposits generally a single egg (although a case has recently come under my notice of three Cuckoo's eggs being found in a Hedge Sparrow's nest containing four of the bird's own production) in the nest of the Meadow Pipit, Tree Pipit, Pied Wagtail, Grey Wagtail,

CUCKOO'S EGG IN PIED WAGTAIL'S NEST.

(THE ONE ON THE LEFT IS THE CUCKOO'S EGG)

Hedge Sparrow, Sedge Warbler, White-throat, Robin, Yellow Hammer, and other small birds. It has been known to deposit its egg in even the nests of such species as Wood Pigeons, Jays, and House Martins, and I am aware of one well-authenticated instance of a Cuckoo's egg in a Carrion Crow's nest. I have proved by a series of experiments with wooden eggs that it is quite easy for the Cuckoo to impose upon the majority of British birds. Such is the passion for brooding that a Starling has been known to hatch out the chick of a common barn-door fowl from an egg substituted for her own clutch.

YOUNG CUCKOO IN MEADOW PIPIT'S NEST.
N.B.—BIRD'S OWN EGGS HAVE BEEN EJECTED.

Our full-page illustration is from a photograph of a Pied Wagtail's nest, containing a Cuckoo's egg, which could only be distinguished by its greater size and rounder shape. The nest was situated about nine feet from the ground, amongst ivy growing over a high garden wall. A common summer visitor to all parts of the British Isles. I have noticed that in the more elevated parts of the north of England Meadow Pipits rear more young Cuckoos than all the other foster-parents put together.

*Materials.*—None.

*Eggs.*—It is certain that the bird lays more than one egg; but although naturalists of good repute have mentioned the number as five, and others have been of opinion that even a larger number may be laid, there is, so far as I know, no reliable evidence to support either supposition. The egg of the Cuckoo is very small in size compared with its layer, remarkably heavy and thick in the shell, and varies very much in coloration, but, strangely enough, often harmonises closely with those of the bird in whose nest it is deposited. It is usually reddish-grey, mottled and spotted closely, with darker markings of the same colour, or pale greyish-green, marked with spots of the same colour. Size about .87 by .75 in. (*See* Plate V.)

*Time.*—April, May, and June.

*Remarks.*—Migratory, arriving in April and leaving in July, the young ones in August and September. Notes: song, *cuckoo*, a bubbling laugh-like note, a churring one, and several others, when the bird is heard at very close quarters, and especially so whilst it is angry. Notes of female resemble bubbling chatter of a dabchick, Local and other names: Gowk, Common Cuckoo. The young Cuckoo turns out all the eggs or chicks that may be in the nest in which it is hatched, an operation I witnessed on one occasion.

YOUNG CUCKOO AND TREE PIPIT FOSTER-MOTHER.

## CURLEW, COMMON.

(*Numenius arquata.*)

Order LIMICOLÆ; Family SCOLOPACIDÆ (SNIPES).

YOUNG CURLEW.

*Description of Parent Birds.*—Length varying from twenty-one to twenty-two inches. Bill very long, slender, curved downwards, and dark brown, paler at the base of the under mandible. Irides hazel. Head, neck, upper part of back, scapulars, and wing-coverts pale brown, with a dark brown streak in the centre of each feather. Wing-quills black, spotted and marked with light brown on the inner webs. Lower back and rump white, marked by a few dusky spots. Upper tail-coverts white, marked with dark brown; tail-feathers barred with dull yellowish-white and dark brown. Chin white; throat and upper part of breast very pale brown, marked with dark brown streaks; lower part of breast, belly, and vent under tail-coverts white, spotted on the two first, with blackish-brown and a dusky streak or two on the latter. Legs long, and, like the toes, bluish-grey in colour.

The female is similar in plumage, but is larger, sometimes even to the extent of five inches in length.

*Situation and Locality.*—On the ground amongst long, coarse grass, tufts of rushes and heath; sometimes quite exposed on bare ground. On rough, undrained pasture lands, moors, and uplands in the

COMMON CURLEW'S NEST AND EGGS.

west and north of England, Wales, Scotland, and Ireland. Our illustrations are from photographs taken on the Westmorland Fells, where these birds are very common. We found a couple of nests within a few yards of each other, the one containing two, and the other three eggs; and the specimen figuring on the previous page was only just over the wall in an adjoining pasture.

*Materials.*—A few short bits of dead rushes, withered grass, or dead leaves, placed in some small declivity; sometimes nothing whatever.

*Eggs.*—Four, sometimes only three, varying from olive-green to brownish-buff in ground-colour, spotted and blotched with dark green and blackish-brown. Size about 2.65 by 1.85 in. (*See* Plate IX.)

*Time.*—April, May, and June.

*Remarks.*—Resident, but resorting to the coast-line during winter. Notes: *curlew*, *curlew*, uttered something like *gurleck*, *gurleck*, when the bird is alarmed. Also a long bubbling kind of whistle in ascending scale during the breeding season. Local and other names: Whaap, or Whaup, Stock Whaap. A very light sitter.

ADULT CURLEW GOING ON TO NEST

## CURLEW, STONE. *Also* NORFOLK PLOVER, GREAT PLOVER, *and* THICKNEE.

(*Œdicnemus scolopax.*)

Order LIMICOLÆ ; Family ŒDICNEMIDÆ (PLOVERS).

*Description of Parent Birds.*—Length about seventeen inches. Bill, short compared with that of the Common Curlew, strong, nearly straight, greenish-yellow at the base, and black at the tip. Irides golden-yellow. A light-coloured streak runs from the base of the beak, under the eye, to the ear-coverts, followed by a brown one running in the same direction below it. Crown, nape, and back of neck light brown, streaked with black. Back, wings (except primaries, which are nearly black, with a white patch on the end of the first and second feathers), and upper tail-coverts light brown, each feather having an elongated blackish-brown centre. Upper half of tail-quills of two shades of brown, producing a mottled effect, followed by a band of white and a black tip. Chin and throat white ; front of neck and breast very light brown, streaked with blackish-brown ; belly and sides nearly white, streaked with brown. Vent and under tail-coverts creamy-white, unmarked. Legs and toes yellow ; claws black.

The female is very much like the male in her plumage.

*Situation and Locality.*—On the ground in warrens, on downs, heaths, and dry commons, principally in Norfolk and Suffolk, but found in several other counties, as far west as Dorsetshire and as far north as Yorkshire. Our illustration was procured on a common in Norfolk.

STONE CURLEW'S NEST AND EGGS.

*Materials.*—Sometimes a few bits of grass, but generally nothing whatever, in the slight declivity made or selected.

*Eggs.*—Two, varying in ground-colour from greyish-yellow to clay-colour, blotched, spotted, and streaked with dark brown, light brown, and greyish-blue. They harmonise very closely with their surroundings. Size about 2.1 by 1.55 in. (*See* Plate VIII.)

*Time.*—May and June. Eggs have, however, been found as late as September.

*Remarks.*—Migratory, arriving in April and departing in October or November. Note: very loud and shrill, and uttered particularly at dusk of evening. Local and other names: Norfolk Plover, Stone Plover, Thicknee, Common Thicknee, Thick-kneed Bustard, Whistling Plover. A light sitter.

---

## DABCHICK. *See* GREBE, LITTLE.

---

## DAW. *See* JACKDAW.

---

## DIPPER. *Also* WATER OUZEL.

(*Cinclus aquaticus.*)

Order PASSERES; Family CINCLIDÆ (DIPPERS).

*Description of Parent Birds.*—Length about seven and a half inches. Bill of medium length, nearly straight, and black. Irides hazel. Head and back of neck dark brown. Back, wings, rump, and tail, which is short, black. Chin, throat, and upper breast snowy white. Belly chestnut brown or rust colour, vent and under tail-coverts black. Legs, toes, and claws black.

*Situation and Locality.*—I have found this bird's nest in niches of rock, on sloping ledges, on a boulder in the middle of a mountain stream, behind a waterfall, in the root of a tree, in the arch of a bridge where a stone had slipped out, fixed to a sod which was constantly dripping with splashes from a waterfall close by, and on one or two occasions in trees some ten or twelve feet over streams. It is never far from a mountain torrent, and is met with in the west and north of England, Wales, and pretty generally over Scotland and Ireland.

DIPPER.

*Materials.*—The exterior is made of aquatic mosses, generally harmonising closely with surrounding objects, and the inside is beautifully lined with dead leaves laid layer upon layer. The appearance of the nest varies considerably according to situation.

*Eggs.*—Four to six, generally five, of a delicate, semi-transparent white, unspotted. Size about 1.0 by .75 in.

*Time.*—March, April, May, June, and occasionally as late as July. I used to notice when a boy that in some seasons the Thrush, and in others the Dipper, would commence to nest first in the north of England.

*Remarks.*—Resident. Call notes: *chit, chit.* Song low and sweet, but very pleasant, and uttered

DIPPER'S NEST

right through the hard frosts of winter during gleams of sunshine. Local and other names: Water Ouzel, Bessy Dooker, Brook Ouzel, Water Crow, Water Piet, Water Crake. Sits closely, and when suddenly disturbed will often dive into any pool that is near.

DIPPERS' HAUNT.

---

## DIVER, BLACK-THROATED.

(*Colymbus arcticus.*)

Order PYGOPODES; Family COLYMBIDÆ (DIVERS).

*Description of Parent Birds.*—Length about twenty-six inches. Bill rather long, straight, pointed, and black. Irides red. Top of head and back of neck grey, darkest on the fore part of head. Sides of neck marked with longitudinal black and white lines. Back scapulars and wing-coverts black, the two first being marked with square patches and the last with round spots of white. Wing-quills, rump and tail-feathers, dusky-black. Chin and throat black, divided by a collar of black and white short longitudinal lines. Breast,

BLACK-THROATED DIVER'S NEST AND EGGS.

belly, and vent white. Under tail-coverts dusky. Legs, toes, and webs dark brown on the outside, reddish or pale brown on the inside.

The female is a trifle smaller than the male, and has a darker head and neck.

*Situation and Locality.*—In a hollow on the ground, amongst the stones and shingle of secluded mountain tarns and loch shores, sometimes amongst

BLACK-THROATED DIVERS' BREEDING HAUNT

the grass, but rarely far from the water (the farthest I have ever seen was ten feet); also on small grassy islands in bodies of fresh water. Indeed, most of the nests I have seen have been situated on small islands. The bird is a great lover of old haunts. In the north-west of Scotland and the Outer Hebrides.

*Materials.*—Roots, stalks, or aquatic herbage, lined with grass, sometimes nothing whatever. The nest figured in our illustration was made entirely of rushes plucked from its immediate surroundings.

BLACK-THROATED DIVER ON NEST.

*Eggs.*—Two, occasionally only one, ranging in colour from buffish-brown to dark olive-brown, scantily spotted with umber and blackish-brown. Average size about 3.25 by 2.0 in. (*See* Plate XIII.)

*Time.*—May and June.

*Remarks.*—Resident, but subject to southern movement in winter. Notes: strange and weird, and said to resemble "*Drink! drink! drink! the lake is nearly dried up.*" Local and other names: Speckled Loon, Lumme, Northern Douker. Does not sit closely. Whilst waiting to secure the photograph of the bird on her nest represented in our photogravure, I had ample opportunities of proving the assertions of older naturalists that the species cannot walk in an upright position. Every time the Diver came back to her nest she pushed her way on her belly from the water to her eggs.

---

## DIVER, RED-THROATED.

(*Colymbus septentrionalis.*)

Order PYGOPODES; Family COLYMBIDÆ (DIVERS).

*Description of Parent Birds.*—Length about twenty-four inches. Bill rather long, straight, sharp-pointed, and bluish horn colour. Irides red. Face and crown ash-grey, nape nearly black, with short perpendicular white lines on it. Back, wings, and upper tail-coverts almost black, spotted with white, except wing-quills, which are uniform black. Chin, cheeks, and sides of neck ash-grey, mixed with lines and spots of a lighter tinge. On the upper part of the neck in front is a conical patch of chestnut-red. Breast, belly, vent, and under tail-coverts white; sides greyish-black, spotted

RED-THROATED DIVER'S NEST AND EGG.

with white. Legs, toes, and webs dark brown on the outside and lighter within.

The female is somewhat smaller in size than the male.

*Situation and Locality.*—On the turf or amongst stones and shingle close to the edges of lonely moorland and mountain pools, tarns, or lakes; by preference on a small island in any of the above sheets of water, on the northern and western mainland, and the islands off those coasts of Scotland. The bird is also said to breed in the west of Ireland.

*Materials.*—Loose rushes and dry grass, very often nothing at all.

*Eggs.*—Two. Olive, or deep greenish-brown in ground-colour, spotted with blackish-brown. Size about 2.8 by 1.8 in. (*See* Plate XIII.)

*Time.*—May and June.

*Remarks.*—Resident, but subject to much local movement. Notes: *kakera, kakera,* uttered during the breeding season. Local and other names: Rain Goose, Kakera, Cobble, Speckled Diver, Spratoon, Sprat-borer. Not a close sitter. I have seen as many as seven members of this species on the wing together in the Outer Hebrides even in May.

---

## DOTTEREL, COMMON.

(*Eudromias morinellus.*)

Order LIMICOLÆ; Family CHARADRIIDÆ (PLOVERS).

*Description of Parent Birds.*—Length about nine and a half inches. Bill of medium length, straight and black. Irides dark brown. Crown and nape blackish-brown. A broad white line runs from the base of the beak, over the eye, down behind the ear-coverts, which are ash-coloured, as also are the neck and back. Wings ash-brown, except quills, which are ash-grey. Tail olive-brown tipped with white; chin white; throat and sides of neck grey. A white gorget-like band, bordered on either side by a dark line, runs across the breast and ends at each shoulder. Breast pale, dull orange; belly black; vent and under tail-coverts white, slightly tinged with buff. Legs and toes greenish-yellow; claws black.

COMMON DOTTEREL GOING ON TO EGGS.

The female is a trifle larger, and more handsomely marked and coloured. It has been stated that the male assists the female in the work of incubation. The bird figured in our illustration appeared to do all the work, and whenever his mate approached within a certain distance he rushed at her like a winged fury. She invariably ducked and escaped the intended blow, but was ultimately chased right away from the neighbourhood of the nest. These observations extended over the best part of two days, and I saw two or three other females by themselves on mountain tops, where I have no doubt they had sitting mates.

*Situation and Locality.*—On the ground amongst woolly-fringe moss, short heather, lichens, and other coarse mountain vegetation, on high, wild moorland districts and lonely mountains of Scotland. It used to breed in the Lake District, but during

COMMON DOTTEREL'S NEST AND EGGS

recent years only a solitary pair or two have been known to attempt to do so, and I fear they have been promptly robbed. I think the greed of the fly-fisher has for ever sealed its doom so far as England is concerned. I am myself an ardent fly-fisher, and have been offered handsome sums of money in my artificial fly-dressing days if I would only procure the skin of this bird, whose feathers are popularly supposed to exercise a kind of charm over trout in some northern districts.

COMMON DOTTEREL ON NEST.

*Materials.*—None as a rule, although a few bits of lichen and dead grass were in the nest figured in our illustration. The eggs are simply deposited in a slight declivity trodden in the place selected.

*Eggs.*—Three, yellowish-olive to dark cream in ground-colour, thickly blotched and spotted with dark brown or brownish-black. Size about 1.65 by 1.15 in. (*See* Plate VII.)

*Time.*—May, June, and July. Although the species breeds late, chicks are sometimes running about by the middle of June, and a Highland keeper on one occasion told me he had found a nest containing fresh eggs on the 26th of July.

*Remarks.*—Migratory, arriving in April and May and leaving in August and September. Notes: *durrdroo*. Local and other names: Foolish Dotterel, Dotterel Plover. Gregarious. Sits fairly closely, and is very tame. Two days' acquaintance with the bird figured put me on such good terms with him that I actually at last stroked his back as he sat on the nest in view of two keepers.

---

DOTTEREL, RINGED. *See* PLOVER, RINGED.

---

## DOVE, RING.

(*Columba palumbus.*)

Order COLUMBÆ; Family COLUMBIDÆ (PIGEONS).

*Description of Parent Birds.*—Length about seventeen inches. Bill moderately long, curved downwards slightly at the tip and pale red, yellowish towards the end and whitish on the soft parts surrounding the nostrils. Irides straw colour. Head and upper part of neck bluish-ash colour. The sides of the lower part of the neck are beautifully glossed with green and purple, according to the light upon them. On either side of the neck is a patch of glossy white, which almost meets at the back. Back and wing-coverts bluish-grey, with exception of a few feathers of the latter, which are white, and during flight form a conspicuous patch, by which the bird may easily be distinguished from any other member of the Pigeon family. Quills dark grey edged with white; the feathers of the spurious wing are almost black. Tail-quills of varying shades of grey, darkest towards the tip.

Chin bluish-grey; neck and breast glossy purple and green; belly and under-parts light ash-grey. Legs and toes red; claws brown.

The female is somewhat duller in plumage and smaller in size.

*Situation and Locality.*—In fir, yew, whitethorn, and various other kinds of trees. I have met with it on the crown of a pollard, and frequently on the ivy-clad trunk of a tree, growing almost at right

RING DOVE.

angles from high rocks and precipices. I have also met with it upon several occasions in an isolated thorn bush growing in the middle of a large field. The nest is situated at a height of from five to seventy or eighty feet, and is found nearly all over the United Kingdom where suitable woodland is to be met with.

*Materials.*—Dead twigs and sticks woven into a loose platform. The nest is often such a poor, flimsy affair that the eggs may be seen through it from beneath, and it is frequently blown down by gales of wind. On the other hand, I have on one or two occasions met with nests of a substantial

RING DOVE'S NEST AND EGGS.

character, into which bits of dried sods had been introduced, so that not even a ray of light could find its way through. Old nests of one season are sometimes repaired and used the following year.

*Eggs.*—Two. White and glossy, similar to those of the Rock Dove, but larger. Average size about 1.65 by 1.25 in.

*Time.*—March, April, May, June, and July. However, nests have been found in nearly every month of the year.

*Remarks.*—Resident. Notes: a soft *ku, ku, ku-ku ku ku ku ku-ku ku ku ku ku-ku kuk.* Local and other names: Cushat, Wood Pigeon, Quest, Cushie Doo, Ring Pigeon. Sits pretty closely when well hidden, but lightly when in an exposed situation.

---

## DOVE, ROCK.

(*Columba livia.*)

Order Columbæ; Family Columbidæ (PIGEONS).

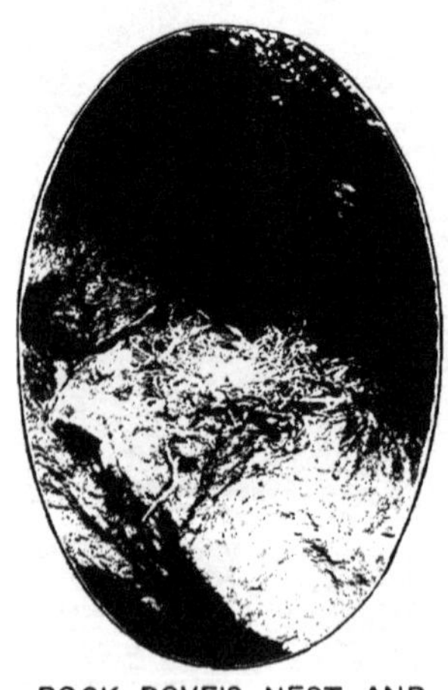

ROCK DOVE'S NEST AND EGGS.

*Description of Parent Birds.*—Length about fourteen inches. Bill of medium length, nearly straight, and brown tinged with red. Irides light reddish-orange. Head and neck dark bluish-grey, glossed with a purply-red and green sheen on the latter; back and wing-coverts light pearl-grey; greater coverts and secondaries marked with black bars; quills bluish-grey, darkish towards the tips; rump white; tail-coverts bluish-grey; quills bluish-grey with a darker bar

ROCK DOVE'S NEST AND EGGS.

at the tips; chin dark bluish-grey; throat and upper part of breast glossed with green, lavender, purple, and purplish-red; lower breast and all under-parts grey. Legs and toes purplish-red; claws brown.

The female is somewhat smaller, and not so brilliant and distinctive in her coloration.

*Situation and Locality.*—Ledges and clefts of maritime and inland cliffs, generally the former, round the coasts of England, Wales, Scotland, and Ireland, wherever suitable accommodation is to be met with. Our illustrations are from photographs taken in a Hebridean cave where traces may be constantly seen of tame pigeons reverting to their original type.

*Materials.*—A small collection of twigs, sticks, seaweed, and bents, roughly constructed, and flat. I have found many nests formed entirely of dead seaweed, and a friend of mine recently had one sent to him from the Outer Hebrides constructed entirely from bits of wire.

*Eggs.*—Two, white, unspotted and smooth. Size about 1.45 by 1.15 in.

*Time.*—March, April, May, and June, although eggs have been found in nearly every other month of the year.

*Remarks.*—Resident. Notes: *coo-roo-coo*, last syllable prolonged. Local and other names: Rockier, Wild Pigeon, Rock Pigeon, Wild Dove, Doo. A fairly close sitter, and distinguished from the Stock Dove by its white rump, which is conspicuously shown as the bird flies out of some sea cave beneath the ornithological student's feet.

## DOVE, STOCK.

(*Columba œnas.*)

Order Columbæ ; Family Columbidæ (PIGEONS).

STOCK DOVE'S NEST AND EGG.

*Description of Parent Birds.*—Length about thirteen and a half inches. Bill of medium length, nearly straight, and pale red, whitish at the tip. Irides brown. Head, neck, and upper parts of back deep bluish-grey, glossed on the sides of the neck with green and purplish-red. Wing-coverts bluish-grey, spotted and marked with black on the greater ones; quills brownish-grey, turning bluer towards the tips. Rump and upper tail-coverts pale bluish-grey. Tail bluish-grey for about two-thirds of its entire length, then crossed by a band of lighter ash-grey, and the end, which is rounded, of so dark a grey that it may almost be described as black; the exterior webs of the outside feathers nearly white. Breast pale reddish-purple; belly, thighs, and under tail-coverts ash-grey; legs and toes red.

The female is a little smaller, and her markings are not so well defined.

*Situation and Locality.*—In hollow trees, rabbit-holes, crevices of rock; in quarries and cliffs, old Crow's or Magpie's nests; in ivy growing against trees, rocks, towers, and steeples; and sometimes quite on the ground, under dense furze bushes. The young birds figured in our illustration were

CLIFF IN WHICH NUMBERS OF STOCK DOVES BREED.

reared in the hay-loft of an old stone barn amongst the Westmorland fells. The second illustration represents a much-frequented cliff in Westmorland during the breeding season. The bird is a common breeder in the eastern and midland counties, and many other parts of England and Wales, and has gained a footing in both Scotland and Ireland, according to Mr. Dixon.

*Materials.*—Twigs, roots, and straws in small quantities, and arranged with very little care or skill.

*Eggs.*—Two, white, faintly tinged with cream colour. They are smaller than those of the Ring Dove, and the creamy tinge distinguishes them from those of the Rock Dove. Size about 1.45 by 1.15 in.

*Time.*—February to October.

*Remarks.*—Resident. Notes: *coo-oo—oo*, the last syllable longer than the first. Local and other names: Stock Pigeon, Wood Dove, Wood Pigeon (a name also used for the Ring Dove). Sits closely. Gregarious, as a rule. It may easily be distinguished from the Rock Dove by its lack of a white rump.

YOUNG STOCK DOVES IN NEST.

## DOVE, TURTLE.

(*Turtur communis.*)

Order COLUMBÆ ; Family COLUMBIDÆ (PIGEONS).

TURTLE DOVE.

*Description of Parent Birds.*—Length about twelve inches. Bill of medium length, slightly curved downward at the tip, and brown. Irides reddish-brown. Crown and back of neck ash-grey, mixed with olive-brown. Back, part of wings, and rump ash-brown, lightest on the margins of the feathers. Wing-quills dusky-brown with lighter margins and tips. Tail-coverts dusky-brown, quills the same in the centre, rest dark grey tipped with white, with which the outside feathers on either side are margined. Chin pale brown ; throat and upper breast light purplish-red, fading into grey. The sides of the neck are marked with a patch of black, each feather of which is tipped with white. Belly, vent, and under tail-coverts white. Under-side of tail-feathers black, deeply tipped with white, except two centre ones, which are of a uniform dusky brown.

The female is rather smaller in size, lacks the black feathers tipped with white on the sides of the neck, and is duller and less distinctive in her coloration.

*Situation and Locality.*—In tall, rough hedges, whitethorn and holly bushes ; in woods, plantations, copses, and spinneys. Common in the southern, midland, and eastern counties ; scarcer in the west and north ; but it is doubtful as to whether

TURTLE DOVE'S NEST AND EGGS.

it breeds in either Scotland or Ireland. I have seen odd specimens in the Outer Hebrides. Our illustration is from a photograph taken in Surrey, where I have found five or six nests in little more than a couple of acres of wood. The flimsy structure is placed at a height varying from four or five to twenty feet from the ground. I have twice found the bird sitting on a Blackbird's nest, the interior of which she had filled up with dead bits of slender birch twigs.

*Materials.*—Sticks and twigs carelessly made into a slight platform, through which the eggs may invariably be seen from beneath. I remember once finding one, amongst some tall ash saplings, that was so slightly constructed with birch twigs as to endanger the eggs slipping through it. I have, however, on the other hand, seen specimens consisting of a thick and bulky platform made entirely of roots of weeds collected from an adjoining ploughed field.

*Eggs.*—Two, creamy-white, glossy, oval, and unspotted. Size about 1.18 by .90 in.

*Time.*—May, June, July, and August.

*Remarks.*—Migratory, arriving in April and May, and departing in September and later. Notes: *tur-tur*, repeated rapidly. They are not unfrequently mistaken for the croaking of a frog by the uninitiated, as I have once or twice discovered by overhearing conversations whilst I was in hiding close to road sides. Local and other names: Wrekin Dove, Ring-necked Turtle, Common Turtle. Sits pretty closely when incubation has advanced, and when disturbed flies off without demonstration.

BLUE-HEADED WAGTAIL. | GREY WAGTAIL. | PIED WAGTAIL. | WHITE WAGTAIL. | YELLOW WAGTAIL.

BEARDED TIT. | BLUE TIT. | COAL TIT. | CRESTED TIT. | GREAT TIT.

MARSH TIT. | LONG-TAILED TIT. | TREE CREEPER. | GOLD-CREST. | TREE PIPIT.

WOODCHAT SHRIKE. | RED-BACKED SHRIKE. | SISKIN. | CHIFFCHAFF. | NUTHATCH.

WHITETHROAT. | LESSER WHITETHROAT. | ST. KILDA WREN. | COMMON WREN. | DARTFORD WARBLER.

NOTE.—*In referring to the eggs the above names should be read from left to right.*

PLATE 3.

| | | | | |
|---|---|---|---|---|
| BLUE-HEADED WAGTAIL. (*See p.* 444.) | GREY WAGTAIL. (*See p.* 448.) | (*See p.* 450.) | (*See p* 452.) | (*See p.* 455.) |
| BEARDED TIT. (*See p.* 422.) | BLUE TIT. (*See p.* 424.) | COAL TIT. (*See p.* 427.) | CRESTED TIT. (*See p.* 431.) | TIT. |
| MARSH TIT. (*See p.* 440.) | LONG-TAILED TIT. (*See p.* 438.) | TREE CREEPER (*See p.* 46) | GOLD-CREST. (*See p.* 128.) | (*See p* 283.) |
| WOODCHAT SHRIKE. (*See p.* 356.) | RED-BACKED SHRIKE. (*See p.* 355.) | SISKIN. (*See p.* 358.) | CHIFFCHAFF. (*See p.* 29.) | NUTHATCH. (*See p.* 239.) |
| WHITETHROAT. (*See p.* 490.) | LESSER WHITETHROAT. (*See p.* 492.) | ST. KILDA WREN. (*See p.* 514.) | | DARTFORD WARBLER. (*See p.* 458.) |

NOTE.—*In referring to the eggs the above names should be read from left to right.*

## DUCK, EIDER.

(*Somateria mollissima.*)

Order ANSERES ; Family ANATIDÆ (DUCKS).

YOUNG EIDER DUCK.

*Description of Parent Birds.*—Length about twenty-five inches. Bill moderately long, thick at the base, straight, and dirty green, whitish at the tip. Irides brown. The top of the head, including the parts round and on a level with the eyes, velvety black, turning to a palish green on the ear-coverts and back of the head. Neck, back, wing-coverts, and scapulars white. Some of the coverts are elongated, and curved at the ends falling over the quills ; greater coverts, and quills black. Rump black, tail-feathers brownish-black. The lower part of the neck is white, tinged with buff. Breast, belly, and under-parts black ; flanks patched with white. Legs, toes, and webs dusky-green ; claws dusky.

The female is somewhat smaller, and her plumage is of pale reddish-brown colour, variegated with brownish-black.

*Situation and Locality.*—On the ground amongst coarse grass, in heather, in clefts of rock, and sometimes on collections of seaweed, amongst shingle, on rocky islands, and on the coast at suitable places round Scotland, in the Orkneys and Shetlands, and at the Farne Islands, the only place on the English coast at which the species now breeds. Owing to protection the birds breeding at the Fàrne Islands are very tame. We stroked the back of one bird as she sat on her eggs close under the walls of St. Cuthbert's Tower. The keepers told us the same bird had nested in that situation for seven

EIDER DUCK'S NEST AND EGGS.

years in succession, as they were able to identify her by a white spot at the back of her head.

*Materials.*—Dry seaweed, heather, or coarse grass, with an inner lining of beautiful soft down from the bird's own body. The down is accumulated as the eggs are laid or incubation advances. Individual birds vary in respect to supplying it during the time they are laying. I have noticed that whilst some had a fairly liberal supply and only three eggs, others with seven had not a particle.

*Eggs.*—Four to eight. Pale greyish-green to

grey cream colour, smooth and unspotted. Size about 3.0 by 2.0 in.

*Time.*—May and June.

*Remarks.*—Resident, with a more southern range in winter. Note: a harsh *kr, kr, kr.* Love note of male: *ah-oo.* Local and other names: St. Cuthbert's Duck, Common Eider, Colk, Dunter Duck. Sits very closely indeed.

EIDER DUCK ON NEST.

---

## DUCK, PINTAIL.

(*Dafila acuta.*)

Order ANSERES; Family ANATIDÆ (DUCKS).

*Description of Parent Birds.*—Length about twenty-seven inches, several of which are accounted for by the abnormally long tail. Bill moderately long, nearly straight, and dusky-black, leaden-grey on the sides. Irides dark brown. Head and upper part of neck in front, nape, and all back of neck dark reddish-brown. The sides and back of head are glossed with purple. Back, wing shoulders, and parts in front of them, grey; wing tertials elongated, black in the centre, and bordered with white and grey; greater coverts ash-brown tipped with reddish-buff and white. Secondaries black, the outer web of each forming a patch of dark green; primaries greyish-brown. Tail-coverts ash-grey, elongated narrow-pointed feathers black, rest dark brown bordered with white. From the side of the head level with the eye a white streak runs down each side of the neck, widening gradually as it descends until the middle of the neck is reached; thence it opens out round the front of the neck, breast, and belly. Sides grey; vent and under tail-coverts black. Legs, toes, and webs blackish-brown.

The female is somewhat smaller, and her plumage is made up of varying shades of brown, the darkest colours in the centre of each feather and on the upper parts of the body. During July, August, and September the male assumes the dress of the female.

*Situation and Locality.*—On the ground amongst grass, rushes, and similar herbage growing near ponds, lakes, and arms of the sea. It is a very rare breeder indeed in our islands, having only been reported from two or three quarters in Ireland and one or two in Scotland and in the Hebrides.

*Materials.*—Reeds, grass, and other kinds of dead vegetation, according to some authorities, lined with brownish tufts of down, faintly tipped with white, from the bird's own body. In four nests which I had the rare good fortune to examine in a single day upon an island which unfortunately must

PINTAIL DUCK'S NEST AND EGGS.

remain nameless, there was very little material beyond dead grass and down.

*Eggs.* — Six to ten, usually seven or eight, smooth, rather elongated, and "greenish-white," according to Mr. Saunders, and "pale buffish-green," according to Mr. Dixon, in colour. The eggs in the four nests which I examined varied from greenish-grey to drab-grey. Size about 2.15 by 1.55 in.

*Time.*—May.

*Remarks.*—Migratory as a rule, wintering with us and spending the summer in Iceland. It is probable that the few staying to breed in our islands have a farther southern range than those from Iceland. Note: a soft inward *quack*. Local and other names: Winter Duck, Cracker, Sea Pheasant, a name, however, more often applied to the Long-tailed Duck. A fairly close sitter, and has bred repeatedly in confinement.

---

## DUCK, TUFTED.

(*Fuligula cristata.*)

Order ANSERES; Family ANATIDÆ (DUCKS).

*Description of Parent Birds.*—Length about seventeen inches. Bill of medium length, but little broadened towards the point, and bluish-grey with a black tip. Irides rich dark yellow. Head and neck glossy purplish-black. The feathers at the back of the head are narrow and elongated, and form a tuft or crest. Back, rump, wings, and tail black with a white bar running across the secondaries, which are tipped with black. There is a small spot of white on the chin. Breast, belly,

TUFTED DUCK'S NEST AND EGGS.

and sides white. Vent and under tail-coverts black. Legs, toes, and webs dusky-black.

The female has all those parts which are black in the male a dusky-brown, and the white parts dirty grey, marked with irregular lines on the sides and flanks. She lacks the crest, or only enjoys it in a very modified form.

*Situation and Locality.*—In a tuft or bush of long, coarse vegetation, such as rushes, sedges, heather, or bent grass, on the edges of tarns and lakes and other suitable places throughout the British Isles. I have met with odd pairs nesting round tarns on the North Yorkshire moors. It nests most numerously in Nottinghamshire. Our illustration was procured on a celebrated Norfolk mere, the nest being in an ideal position, the bird having not more than twelve or fourteen inches to travel from the edge of her nest into deep water.

*Materials.*—Rushes, sedges, reeds, and grass, with an inner lining of down tufts plucked from the bird's own body. These tufts are greyish-black, smaller and a trifle darker than those of the Pochard, with more obscure white centres.

*Eggs.*—Eight to fourteen, usually nine or ten. I have seen one nest with fifteen eggs and another with nineteen. The latter was, in all probability, the work of two females. Pale buff tinged with green. Very similar to those of the Pochard. Size about 2.3 by 1.6 in.

*Time.*—May and June.

*Remarks.*—A winter visitor, though numbers stay to breed. Notes: call, *currugh, currugh,* uttered on alighting. Local or other name: Tufted Pochard. Sits closely.

## DUCK, WILD. *Also* MALLARD.

(*Anas boscas.*)

Order ANSERES; Family ANATIDÆ (DUCKS).

WILD DUCK ON NEST.

*Description of Parent Birds.*—Length about twenty-four inches. Bill of medium length, broad, and yellowish-green. Irides hazel. Head and upper half of neck rich glossy green, below which is a narrow collar of white, succeeded by greyish chestnut-brown; back brown; wings ash-brown, with a broad transverse bar of reflecting purplish- or violet-blue, bounded on either side by a narrow bar of rich black, and another beyond of white. Rump, upper tail-coverts, and four middle tail-feathers, which are curled upwards, rich velvet-black, the rest ash-grey edged with white. Upper part of breast rich dark chestnut; lower breast, belly, and vent greyish-white; under tail-coverts rich black. Legs, toes, and webs orange-yellow.

The female is about two inches shorter, and her plumage is nearly all composed of sober brown and black. She retains the rich bar of violet- or purplish-blue on her wings, but lacks the curled feathers in the tail of the male. About the end of May the male commences to cast his curled tail-feathers and to change his plumage, and during June he assumes the sober female garb, which he wears through July. This he begins to discard in August, and between the first and second weeks in October he has again donned his magnificent dress.

*Situation and Locality.*—On the ground amongst rushes, brambles, long rank grass, sedge tufts, under a bunch of heather, and in corn fields and hedge bottoms. Generally near lakes, rivers, tarns, and ponds, or in marshes, bogs, and swamps. They are, however, often found at considerable heights, in faggot stacks, deserted Crows' nests, squirrel dreys, Hawks' nests, in hollow trees, pollards, and other elevated situations in ruins and rocks, from

MALLARD AND WILD DUCKS ON ICE.

which heights the female has been said to convey her progeny upon her back. Pretty general in all suitable places throughout our Isles. In 1902 a case came under my notice of a Wild Duck and Corncrake laying in the same nest in North Uist.

*Materials.*—Dry grass, bracken, or other suitable vegetation near at hand, with a lining of down from the bird's own body. The tufts are neutral grey, tipped very slightly, with white.

*Eggs.*—Eight to fifteen or sixteen, generally ten to twelve; greenish-white tinged with buff. Size about 2.3 by 1.6 in.

*Time.*—February, March, April, May, June, and even as late as November, individual nests have

WILD DUCK'S NEST AND EGGS.

been recorded. April and May are the principal months, however.

*Remarks.*—Resident, and partially migratory, larger numbers visiting us in winter than stay to breed in summer. Notes: *quack*, loud and high sounding when uttered by the female, and harsh and low when by the male. Local and other names: Mallard (a name, strictly speaking, applying to the male only), Stock Duck, Common Wild Duck. Sits closely, and covers over her eggs when voluntarily leaving her nest.

---

## DUNLIN.

(*Tringa alpina.*)

Order LIMICOLÆ; Family SCOLOPACIDÆ (SNIPES).

DUNLIN ON NEST.

*Description of Parent Birds.*—Length about eight inches. Beak rather long, nearly straight, and black. Irides brown. Crown and upper parts reddish-brown, streaked on the head and back of neck with dusky-black, and each feather on the back black in the centre. Wing-coverts greyish-brown edged with light grey; quills dusky-black, inclining to brown on some of the lesser, which are greyish-white on the edges of the outer webs. Upper tail-coverts white, quills ashy-brown edged with grey, excepting the two centre feathers, which are longer than the rest and dusky-brown. Chin white, cheeks, throat, and sides of neck and breast whitish, streaked with dusky-black; belly and under-parts

DUNLIN'S NEST AND EGGS.

white. Legs, toes, and claws dusky-black, slightly tinted with green.

The female differs little from the male, but is, as a rule, slightly larger, and has a deeper call-note.

*Situation and Locality.*—On the ground, well hidden by a tuft of ling, heather, or tussock of coarse grass, in boggy, marshy, tarn-besprinkled parts of moors and heaths in the north and ex-

YOUNG DUNLINS HIDING.

treme west of England; also in Scotland and Ireland.

*Materials.*—A few straws or bents forming a slight lining to the hollow in which the eggs are deposited.

*Eggs.*—Four, pear-shaped; ground-colour varies from greenish-white to cream or buff of different shades, blotched and spotted with reddish- and blackish-brown, and underlying markings of grey. Size about 1.3 by .95 in. (*See* Plate IX.)

*Time.*—May and June.

*Remarks.*—Resident, but migratory, partially and locally—that is to say, more birds visit our coasts in winter than stay to breed, and those that

WOOD WARBLER.
(*See p.* 479.)

GARDEN WARBLER
(*See p.* 461.)

MARSH WA
(*See p.* 466

BLACKCAP.
(*See p.* 7.)

GRASSHOPPER WARBLER.
(*See p.* 463 )

WILLOW WA
(*See p.* 4

MISSEL THRUSH.
(*See p.* 416.)

RING OUZEL.
(*See p.* 245.)

STONECHAT.
(*See p.* 388.)

SONG THRUSH.
(*See p.* 419 )

WHEATEAR.
(*See p.* 480.)

ROBIN.
(*See p.* 328.)

WHINCH
(*See p.* 48

NOTE.—*In referring to the eggs the above names should*

PLATE 4.

do breed with us resort to the coast-line in winter. Notes: call, *kwee-kwee, trui*, or *pe, pe, pe*. Local and other names: Dunlin Sandpiper, Purre, Judcock, Stint (the name of a different species altogether), Oxbird, Plover's Page, Churr, Sea Snipe, Sea Lark, Least Snipe. Sits pretty closely.

---

## EAGLE, GOLDEN.

(*Aquila chrysaëtus.*)

Order ACCIPITRES; Family FALCONIDÆ (FALCONS).

*Description of Parent Birds.*—Length about thirty-six inches. Beak moderately long, much curved at the tip, and bluish horn colour; bare skin round the base yellow. Irides hazel. The whole of the plumage is brown; the head, back of neck, and some of the wing-coverts reddish; wing-quills blackish-brown; tail-quills of two shades of brown, darkest at the tip. Chin and throat dark brown; under-parts of the body and thighs bay. The legs are feathered down to the feet, which characteristic distinguishes this bird from the Sea Eagle. The feet are yellow and the claws black. Mr. Booth was of opinion that the Golden Eagle does not assume the full mature plumage until it is five or six years old.

The female resembles the male in plumage, but is somewhat larger in size.

*Situation and Locality.*—On ledges of high inaccessible cliffs and precipices and trees in the wildest and most desolate parts of Scotland and Ireland. In some of the Highland deer forests this noble bird is now strictly preserved, and such most commendable hospitality will no doubt save it to us for some time to come. Our illustrations are from photographs taken in the Western Isles of Scotland, but in the interests of British ornithology I think it best not to advertise the exact spot. The pictures were secured at different seasons; when the young ones were in the nest it contained

GOLDEN EAGLE'S NEST AND EGGS.

two partly consumed mountain hares, off which the down had nearly all been carefully plucked, and the hind legs of a half-grown black rabbit.

*Materials.*—Sticks, bits of heather, dead fern-fronds, grass, and moss. The nest is repaired from year to year, and often becomes a bulky structure, on account of the bird using the same site for a long period. The one figured opposite contained a large quantity of sticks and rubbish.

*Eggs.*—Two, sometimes three; very rarely four. Subject to variation both in ground-colour and markings. The commonest type is dingy white,

GOLDEN EAGLE'S EYRIE, WITH YOUNG, AND PARTLY DEVOURED PREY.

clouded, blotched, and spotted nearly all over with rusty or reddish-brown, and underlying markings of grey. Some specimens are pure white, unspotted. Size about 2.9 by 2.35 in. (*See* Plate VI.)

*Time.*—March and April.

*Remarks.*—Resident but wandering. Note: "a barking cry" according to Seebohm; and "a loud yelp uttered several times in succession" according to Dresser and Sharpe. Local or other name: Ring-tailed Eagle, from the fact that young specimens have the basal half of the tail white. Sits fairly closely.

---

## EAGLE, WHITE-TAILED. *Also* SEA EAGLE.

(*Haliaëtus albicilla.*)

Order ACCIPITRES; Family FALCONIDÆ (FALCONS).

*Description of Parent Birds.*—Length about thirty-three inches. Beak somewhat lengthened and nearly straight, except at the tip, where the upper mandible is much hooked. It is very strong, horn colour at the tip, and yellow at the base, as is also the bare skin surrounding that part. Irides very light yellow. Head and neck ash-brown, varying in hue with age; back and wings dark brown, a few lighter coloured feathers being intermingled; wing-quills dusky-black. Tail white. Breast and under-parts dark brown. Legs and toes yellow; claws black.

The female is thirty-eight inches, and both sexes vary greatly in the colour of their plumage.

*Situation and Locality.*—On ledges and in holes of high inaccessible cliffs, generally near the sea; in a tree or upon the ground on a small rocky island in the middle of a mountain loch. In the

WHITE-TAILED EAGLE'S NESTING SITE.

Shetlands, on the west coast of Scotland and the surrounding islands, and in Ireland. A century ago this species bred both in the Isle of Man and the north of England, but it is now growing very rare. It is on the verge of extinction in Ireland.

*Materials.*—Sticks, twigs, seaweed, heather, grass, and wool. The nest is often a huge structure, from the fact that the bird adds to it year by year. Specimens have been known measuring as much as five feet across. It is very shallow.

*Eggs.*—Two generally, sometimes only one; and upon exceptional occasions three have been found. White, usually quite unspotted, but upon rare occasions specimens have been taken slightly marked with pale red. Average size about 3.0 by 2.25 in.

*Time.*—March, April, and May.

*Remarks.*—Resident, but wandering. Note: a yelping, or barking kind of cry. Local and other names: Sea Eagle, Erne, Cinerous Eagle. Sits rather lightly, and is much attached to nesting site.

---

## FALCON, PEREGRINE.

(*Falco peregrinus.*)

Order Accipitres; Family Falconidæ (FALCONS).

PEREGRINE FALCON

*Description of Parent Birds.*—Length from fifteen to seventeen or eighteen inches. Bill short, strong, much curved, and blue with a blackish tip. Bare skin round the base of the beak and eyelids yellow. Irides dusky. Head, back of neck, and upper parts generally bluish-ash, coloured darkest on the crown and nape, and faintly barred on the back and wing-coverts with a darker tint. Wing-quills dusky, barred and spotted on the inner webs with reddish-white. Tail-feathers barred alternately with black and dingy ash. Chin, throat, and upper breast white, tinged with yellow or rufous, and marked on the two latter parts with a few dark streaks. Lower breast, belly, and under-parts white, barred with dark brown and grey. Legs and toes yellow; claws black.

The female is somewhat larger, but similar in plumage. However, the species is subject to a great amount of individual variation.

*Situation and Locality.*—On ledges and in the crevices of rugged inaccessible sea cliffs and inland crags. In a few places in England and Wales, and more numerously in Scotland and Ireland. I know a scaur in Westmorland where the bird frequently attempts to breed, but invariably gets shot or robbed. In almost every great sea-fowl haunt, such as Ailsa Craig, the Bass Rock, and St. Kilda, will be found a pair of Peregrines breeding amongst Guillemots, Razorbills, and Kittiwakes.

*Materials.*—Sticks, dry seaweed, heather, and wool, or hair, bones, and castings, according to some authorities. Our friend Mr. Ussher, who has probably examined more breeding ledges than any other man in the United Kingdom, says in his "Birds of Ireland" that he has "never found any building materials whatever brought by a Peregrine, though she will sometimes lay in the deserted nest of a Raven or Hoody Crow." And our experience coincides with this absolutely. In Mull my brother once found a Peregrine occupying an old Raven's nest, the rightful owners having been driven out

PEREGRINE FALCON'S EGGS.

and forced to build a new home at some distance away, but in the same cliff.

*Eggs.*—Two to four. Morris says on rare occasions even five. Ground-colour varies from light orange-yellow to pale russet-red, thickly spotted, clouded, and mottled with reddish-brown of various shades. Size about 2.05 by 1.6 in. (*See* Plate V.)

*Time.*—April and May.

*Remarks.*—Resident. Note: a loud chatter, sounding like "hec, hec, hec," which is generally uttered when the bird has been disturbed, and is circling high in the air above her breeding haunt. Local or other name: none. Sits lightly or closely, according to position, and is particularly partial to an old nesting site.

In spite of the fact that this noble bird is terribly persecuted for its eggs or young ones in England and Wales, it manages to hold its own in Scotland, although trapped and shot to some extent by keepers.

YOUNG PEREGRINE FALCONS.

## FLYCATCHER, PIED.

(*Muscicapa atricapilla.*)

Order. PASSERES ; Family MUSCICAPIDÆ (FLYCATCHERS).

*Description of Parent Birds.*—Length about five inches. Bill rather short, straight, pointed, and black. Irides dark brown. On the forehead is a small white patch, which differs in point of size in individuals; crown and nape brownish-

PIED FLYCATCHER'S NEST AND EGGS.

black; back black; wing-coverts and quills blackish-brown; edges of greater coverts and outer webs of tertials white. Tail dusky-black, parts of outer and second feathers white. All the under-parts are white. Legs, toes, and claws black.

The female lacks the white forehead, and is generally less distinctive in her coloration.

*Situation and Locality.*—In holes of trees and walls and crevices of rocks, in wild, out-of-the-way parts in the six northern counties of England, Wales, and the south of Scotland. Our photographs

PIED FLYCATCHER'S NESTING SITE.

were secured in Wales, where I have located as many as four nests in a wood of only a few acres. The picture of male and female outside a nesting hole was secured whilst the former was trying to induce the latter to go back to her nest and eggs, near which the terrifying camera stood. When a Pied Flycatcher cannot induce his mate to go back to her maternal duties by coaxing her with food he chases her from tree to tree until he finally compels her to enter the nesting hole. I

PIED FLYCATCHERS AT NESTING HOLE.

have met with the bird once in Essex, but do not think it was breeding.

*Materials.*—Dry grass, moss, leaves, feathers, and hair, loosely put together.

*Eggs.*—Five to eight, generally five or six. Of a uniform pale blue or greenish-blue, closely resembling those of the Redstart, but they are occasionally marked with a few reddish-brown spots, it is said; however, I have never seen any so marked. Size about .75 by .55 in. (*See* Plate V.)

*Time.*—May and June.

*Remarks.*—Migratory; arriving in April and

leaving in September or October. Notes very like those of the Redstart. Local or other name: Coldfinch. A close sitter.

---

## FLYCATCHER, SPOTTED.

(*Muscicapa grisola.*)

Order PASSERES; Family MUSCICAPIDÆ (FLYCATCHERS).

YOUNG SPOTTED FLYCATCHER.

*Description of Parent Birds.*—Length about six inches. Bill of medium length, straight, broad at the base, and dusky-black in colour. Irides dark brown. Head, back of neck, back, rump, and upper tail-coverts brown, the head being spotted with a darker tinge of the same colour. Wings brown, tail the same colour, and a trifle lighter at the tip. Chin, throat, breast, and under-parts a dull white, streaked on the throat and breast with dusky-brown. Legs, toes, and claws dusky-black.

The female is very similar to the male.

*Situation and Locality.*—On the horizontal branches of fruit trees trained against walls, in trellis-work, rose trees trained against houses, in holes in walls, ivy climbing up a wall or the trunk of a tree (as in our illustration), on ledges of rock, on the ends of beams projecting from old houses and sheltered by an overhanging gable or roof, and in almost every conceivable situation. I have seen it inside half of the shell of a Cocoanut, in an old boot thrown into a tree, and other odd sites.

*Materials.*—These vary as considerably as the

SPOTTED FLYCATCHER'S NEST

positions selected for their accommodation. Straws, fibrous roots, moss, hair, feathers, rabbits' down, and cobwebs, somewhat loosely put together, as a rule, but occasionally I have come across a very compact little structure.

*Eggs.*—Four to six, generally five, varying considerably in coloration. The ground-colour ranges from grey to light green, the markings running through various shades of faint red or reddish-brown. Sometimes they are almost entirely absent, at others they form a belt round the larger end, and I have met with eggs with large, bright rust-red spots thickly distributed over the entire surface. Size about .75 by .57 in. (*See* Plate V.)

*Time.*—May, June, and July.

*Remarks.*—Migratory, arriving in the early part of May and leaving in September and October. Notes: a weak chirp and a harsh call-note. Local and other names: Beam-bird, Rafter, Bee-bird, Chanchider, Cherry-sucker, Bee-eater, Post-bird, Cherry-chopper. Sits closely, and flies away without demonstration when disturbed.

SPOTTED FLYCATCHER AT NEST

## GADWALL.

(*Anas streperus.*)

Order ANSERES; Family ANATIDÆ (DUCKS).

*Description of Parent Birds.*—Length about twenty-one inches. Bill of medium length, broad, flat, and leaden-coloured. Irides hazel. Head and upper portion of neck pale brown, mottled with a darker tinge of the same colour; back grey, of two shades running in alternate curved lines; rump and upper tail-coverts bluish-black; wings long and pointed, small coverts reddish-brown, greater nearly black; secondaries brownish-grey, with a conspicuous white patch on them; primaries brown; tail-quills darkish brown, bordered with a lighter tinge of the same colour; lower half of neck dark grey, marked with short curving lines of a lighter tinge; breast and belly white; sides, flanks, and vent marked with irregular vertical lines of two shades of grey; lower tail-coverts black. Legs, toes, and webs dull orange; claws black.

The female has the head and upper part of the neck pale brown, spotted with dark brown; back of neck, back, and rump brown, the feathers being edged with pale reddish-brown; the wings are similar in markings to those of the male, but not so bright; lower part of neck, in front, and breast pale brown, with broad curved bands of dark brown.

*Situation and Locality.*—On the ground amongst reeds, sedges, rushes, and long, coarse grass on small islands situated in lakes; on the banks of broads, pools, in marshes and swamps in Norfolk only, so far as is known. Our illustration was obtained in that county.

GADWALL'S NEST AND EGGS.

*Materials.*—Dead grass, sedges, or leaves, lined with down of a brownish-grey colour, obscurely tipped with white. The tufts are smaller than those found in the Mallard's nest.

*Eggs.*—Five to thirteen, generally from eight to ten. Buffish-white and polished, closely resembling those of other members of the Duck family. Size about 2.1 by 1.5 in.

*Time.*—May and June.

*Remarks.*—At one time only a rare winter visitor, but now a resident with us. Some fifty odd years ago a pair captured in a Norfolk decoy were pinioned and turned loose. They bred and multiplied and, it is thought, induced migrants to stay and do the same, until now there is a very respectable number in the county above mentioned. Notes: a low quacking. Local and other names: Gadwall Duck, Grey Duck, Common Gadwall, Rodge. Sits closely.

---

## GANNET. *Also* SOLAN GOOSE.

(*Sula bassana.*)

Order STEGANOPODES; Family PELECANIDÆ (PELICANS).

*Description of Parent Birds.*—Length about thirty-four inches. Bill about six, straight, broad at the base, and horny greyish-white in colour. Irides pale straw colour. Skin of face and throat bare and blue. Head and neck buff. The whole of the body white, except wing primaries, which are black. The tail is tapering and pointed. Legs, toes, and webs black, with a pea-green line running up each toe and continued up the front of the shank.

The female closely resembles the male. Some authorities say that the bird does not don its adult

GANNETS ON NESTS.

plumage until it is three years old; others place the age limit at four.

*Situation and Locality.*—On the shelves and ledges of precipitous sea cliffs and rocks. The birds breed in colonies, and engage every available situation capable of accommodating a nest. On the Bass Rock, Ailsa Craig, St. Kilda, and other suitable places on the islands lying to the north and west of Scotland; and in one or two places off the Welsh and Irish coasts. Wilson, the American ornithologist, after visiting the above-named breeding haunts, computed that the birds nesting upon them killed more herrings in the course of a year than all the fishermen in Scotland.

GANNET AND YOUNG.

*Materials.* — Seaweed, bits of turf, moss, and grass, sometimes in large quantities. Like rooks, these birds frequently steal materials from each other's nests during the absence of the owners in search of more seaweed, moss, or grass; and I have on several occasions seen great battles waged as a result of the discovery of a thief.

*Egg.*—One. White or bluish-white, covered like that of the Cormorant, with a thick coat of lime, which quickly becomes soiled and dirty by being trodden upon. Size about 3 by 2 inches.

*Time.*—May and June.

*Remarks.*—Resident, but subject to much local

movement. Notes loud and harsh. Local and other names: Soland Goose, Common Gannet, Solan Gannet. Gregarious, and sits very closely.

---

## GARGANEY.

(*Querquedula circia.*)

Order ANSERES; Family ANATIDÆ (DUCKS).

*Description of Parent Birds.*—Length about sixteen inches. Bill fairly long, straight, and black. Irides light hazel. Crown and back of head dark brown, which colour passes down the back of the neck, ending about the middle in a point. Back dark brown, the feathers being bordered with a lighter tinge of the same colour. Wing-coverts ashy-grey, scapulars elongated and narrow, white in the centre, and black round the edges; the transverse reflecting patch on the secondaries is green, bordered with white; primaries brownish-black; tertials grey; tail greyish-brown. A white stripe commences in front of the eye, passes over it and the ear-coverts, and becoming narrower, runs down the side of the neck for some distance. Cheeks and sides of neck reddish-brown, interspersed with fine lines of white pointing downwards. Chin black; throat and breast dark brown, marked with short, semicircular lines of light brown. Belly white; sides and flanks crossed with wavy black lines, which terminate towards the vent in two wide bands. Vent and under tail-coverts mottled with dusky-black. Legs, toes, and webs greyish-brown.

The female differs considerably from the male. She is smaller in size; her head is brown, marked with lines and spots of a darker tinge; back and wing feathers closest thereto dark brown, bordered with rusty brown, and tipped with white; wing-coverts greyish-brown, and green patch on wing duller. The white band over the eye is duller, and tinged with yellow. Chin white; breast greyish-white, marked with two shades of brown. Sides and flanks light brown, marked with a darker tinge of the same colour.

*Situation and Locality.*—On the ground in a tuft

HOME OF THE GARGANEY.

of rushes or sedge; amongst reed beds and coarse rank herbage on the rough banks of broads, rivers, and marshy pools in Norfolk and Suffolk, where alone the bird is now said to breed, and is, unhappily, on the decrease.

*Materials.*—Rushes, leaves, dry grass, and small brown tufts of down with long white tips from the bird's own body.

*Eggs.*—Eight to thirteen or fourteen. Creamy white, of varying shades, very similar indeed to

GARGANEY'S NEST AND EGGS.

those of the Teal, but perhaps a trifle more creamy in tint. Size about 1.8 by 1.35 in.

*Time.*—April and May.

*Remarks.*— Migratory, arriving in February and March, and departing in November. Note: a loud, harsh *knack*. The male has a peculiar, rattle-like note in the spring. Local and other names: Garganey Teal, Garganey Duck, Summer Duck, Summer Teal, Cricket Teal, Crick, Pied Wigeon. Sits very closely.

---

## GOATSUCKER. *See* NIGHTJAR.

---

## GOLD-CREST. *Also* GOLDEN-CRESTED WREN.

(*Regulus cristatus.*)

Order PASSERES; Family SYLVIIDÆ (WARBLERS).

GOLD-CREST.

*Description of Parent Birds.*—Length about three and a half inches. Bill rather short, straight, slender, and black. Irides hazel. Forehead and round the eyes whitish, tinged with dull olive-green. Crown pale orange in front, and darker and richer towards the hind part. The feathers are somewhat elongated, and form a kind of crest, which is bounded on either side by a black streak. Neck, back, rump, and upper tail-coverts olive-green. Wing-quills dusky black, edged with greenish-yellow; coverts black, tipped with white, forming two white bars on wings, plainly visible during flight. Tail-quills dusky, edged with

GOLD-CREST'S NEST

yellowish-green. All the under-parts are greyish-white, tinged with buff on the throat, breast, and sides. Legs, toes, and claws brown.

The female is less distinct in coloration, and her crest is somewhat modified in size.

*Situation and Locality.*—Usually suspended from the branch or branches of a spruce fir; sometimes a cedar, yew, or holly is selected. It is placed near the end of a horizontal branch, at a height varying from two or three to ten or twelve feet from the ground; rarely in bushes, although I have met with it in furze; in woods, plantations, spinneys, shrubberies, and small clumps of trees, pretty generally throughout the United Kingdom where suitable trees are plentiful. Our full-page illustration is from a photograph taken on the outskirts of a large plantation in Norfolk.

*Materials.*—Green moss, lichens, fine grass, spider-webs, caterpillar cocoons, and hair, beautifully felted together, and lined with down and feathers. It is a wonderfully compact little structure, for which its builder has been known to steal materials from the nest of a Chaffinch close by.

*Eggs.*—Four to ten; generally six or seven. Pale flesh colour, or very faint brown; occasionally white, spotted, and suffused, at the larger end generally, with light reddish-brown. Size about .56 by .42 in. (*See* Plate III.)

*Time.*—March, April, May, and June.

*Remarks.*—Resident, and a winter visitor. Notes: song, weak but pleasant; call, a shrill *tsit, tsit.* Local and other names: Golden-crested Wren, Golden-crowned Knight, Golden-crested Warbler, Gold-crested Wren, Gold-crowned Wren, "Woodcock Pilot "—from the fact that, as a winter visitor, it precedes that bird by a few days. A close sitter, and the smallest British bird.

---

## GOLDEN EYE.

(*Clangula glaucion.*)

Order ANSERES ; Family ANATIDÆ (DUCKS).

This species is said to have bred in the north of Scotland, but no reliable ornithological authority has yet verified the statement, so far as I can gather.

---

## GOLDFINCH.

(*Carduelis elegans.*)

Order PASSERES ; Family FRINGILLIDÆ (FINCHES)

*Description of Parent Birds.*—Length about five inches. Bill rather short, nearly conical, whitish at the base, and black at the tip. Forehead and chin rich scarlet, divided by a line of black, which passes from the base of the beak to the eyes. Cheeks white. Crown and back of head black, which colour descends on either side of the neck in a narrowing band. Back and rump pale tawny-brown, lightest on the back of the neck. Wing-coverts black ; quills black, barred across with yellow, and tipped with white. Upper tail-coverts grey, mixed with tawny-brown ; quills black, marked with white and buffy-white spots near their tips. Throat and under-parts white, tinged on the breast, sides, and flanks with pale tawny-brown. Legs and toes pale pinkish-white ; claws brown.

The female is somewhat similar, but is said to have a more slender beak, the red on her head to be less in area, and often speckled with black, and the smaller coverts of the wings to be dusky-brown instead of black.

*Situation and Locality.*—In the fork of an apple, pear, or other fruit-tree in gardens and orchards; on the boughs of chestnut and sycamore trees, evergreens, and sometimes in thick hedgerows. Sparingly throughout England, in some parts of Scotland, and widely, though not numerously, in Ireland.

*Materials.*—Moss, fine roots, dry grass straws, bits of wool, lichens, and spiders' webs, lined with feathers, willow down, and hairs. It is a neat little cup-shaped structure, the materials of which depend to some extent upon what the bird may find lying around.

*Eggs.*—Four to six, greyish, or greenish-white, spotted and streaked with light purplish and reddish-brown and grey. The markings are, as a rule, most numerous round the larger end. The eggs of this species generally run smaller in size than those of the Greenfinch and Linnet, but closely resemble them in colour. Size about .66 by .51 in. (*See* Plate II.)

*Time.*—May, June, and July.

*Remarks.*—Migratory and resident. Notes: call, *ziflit*, or *sticklit;* song: shrill twittering and warbling, and containing the syllable *fink*. Local and other names: Gold Spink, Draw-water, Thistle Finch, Grey Kate, or Pate, Goldie, King Harry, Redcap, Proud Tail. Sits pretty close, and flies away without demonstration.

GOLDFINCH'S NEST AND EGGS.

## GOOSANDER.

(*Mergus merganser.*)

Order Anseres ; Family Anatidæ (DUCKS).

*Description of Parent Birds.*—Length twenty-six and a half inches. Bill rather long, straight, hooked at the tip, and vermilion-red, except the upper ridge and point of the upper mandible, which are black. Irides red. Head and upper half of neck rich glossy green ; feathers on the back of the head lengthened. Upper back and scapulars black ; lower back, upper tail-coverts, and tail-quills ash-grey. Shoulder of wing, all the coverts and secondaries white ; primaries almost black. Lower part of neck in front, breast, belly, vent, and under tail-coverts salmon-buff. Legs and toes orange-red, webs somewhat darker.

The female is rather smaller, and differs to a considerable extent in coloration. Bill and irides duller. Head and upper part of neck reddish-brown. Back, wings, tail-coverts, tail-quills, sides and flanks ash-grey, except secondaries and primaries, which are white and lead-grey respectively. Throat white, breast, and under-parts tinted with buff. Legs and feet orange-red.

*Situation and Locality.*—Holes in trees, clefts in rocks, holes amongst the exposed roots of trees, on ledges of rock, under the cover of bushes, on small islands, in freshwater lochs, on the banks of streams and lochs, in forests in the northern Highlands. No absolute proof of the bird's nesting in the British Isles was forthcoming until as late as 1871.

*Materials.*—Depend somewhat upon position ;

GOOSANDER'S NEST AND EGGS.

those in trees are said to have none except the decayed wood, and down from the bird's own body, whilst in other situations liberal quantities of dead weeds, dry grass, and small roots and down are used.

*Eggs.*—Six to twelve or thirteen. Creamy-white. Size about 2.7 by 1.85 in.

*Time.*—April and May.

*Remarks.*—Migratory, being principally a winter visitor, but a few remaining to breed. Note: a low plaintive whistle. Local and other names: *D*un *D*iver, Saw Bill, Jacksaw, Sparling Fowl. Sits closely.

---

## GOOSE, GREY-LAG. *Also* WILD GOOSE.

(*Anser cinereus.*)

Order ANSERES; Family ANATIDÆ (DUCKS).

GREY-LAG GOOSE ON NEST.

*Description of Parent Birds.*—Length thirty-five inches. Bill of medium length, fairly straight, and pink flesh-colour, except on the tip of each mandible, where it is white; irides brown; head, back of neck, and upper portion of back, ash-brown, the feathers of the last bordered with a lighter tinge; wings lead-grey on the portions nearest the back, each feather being broadly margined with lighter grey, the outer front portion pale bluish-grey, the rest dark leaden-grey; lower portion of the back and rump light bluish-grey; upper tail-coverts white; tail-quills white on the inside webs and

GREY-LAG GOOSE'S NEST AND EGGS.

greyish-brown on the outer, and tipped with white; chin, throat, and breast light grey; belly, vent, under tail-coverts, and under side of tail-quills, white. Sides, flanks, and thighs barred with two

GREY-LAG GOOSE'S NEST.
(*Eggs covered with down.*)

shades of grey; legs, toes, and webs flesh colour; claws black.

The female is smaller in size than the male.

*Situation and Locality.*—On the ground amongst tall rank grass, heather, rushes or osiers in lonely swamps and moorland bogs of Ross, Sutherland, Caithness, and the Outer and Inner Hebrides; also in a semi-domesticated state at Castle Coole, in Ireland, where the birds are only subject to a limited local movement. The bird appears to be partial to small islands, whereon I have seen at least half a dozen nests in a single day.

*Materials.*—Heather, dried flags, rushes, leaves, and grass with an inner lining of feathers and down from the breast of the female.

GREAT CRESTED GREBE ON NEST.

*Eggs.*—Five to nine, the former number being an average clutch according to my experience. Sometimes twelve to fourteen are said to have been found; but all the best authorities have had to accept the latter numbers on hearsay. Dull yellowish, or creamy-white, with a very slight suggestion of green. Size about 3.4 by 2.35 in.

*Time.*—March, April, and May.

*Remarks.*—Resident, but subject to much local movement, and numbers increased by northern arrivals during winter. Note: a "gaggle." Local and other names: Wild Goose, Grey Goose, Grey-legged Goose. Sits closely, and covers eggs with down when voluntarily leaving nest.

GREY-LAG GOSLINGS.

---

## GOSHAWK.

(*Astur palumbarius.*)

Order ACCIPITRES; Family FALCONIDÆ (FALCONS).

Has now quite ceased to breed within the British Isles, and is only a straggler seen upon rare occasions.

## GREBE, GREAT CRESTED.

(*Podicipes cristatus.*)

Order PYGOPODES ; Family PODICIPEDIDÆ (GREBES).

*Description of Parent Birds.*—Length about twenty-two inches. Bill rather long, straight, pointed, black at the tip, and reddish towards the base. The top of the head and the divided crest with which it is adorned are dusky ; cheeks whitish. Round the upper part of the neck is a tippet or ruff, which is formed of elongated feathers that stand out all round. These feathers are rusty red, with a darker tinge at the tip of each. Hind part of the neck, back, wings, and short, tufty tail, dark brown, except the secondaries of the wings, which are white. Front of neck, breast, and belly silvery white. Sides and flanks, pale chestnut ; outside of legs and toes, dusky green ; inside, pale yellowish-green. Each toe is surrounded by a margin of web.

The female is not so large or distinct in coloration. Her crest is also smaller.

*Situation and Locality.*—Amongst reeds growing in the water. Sometimes its foundation rests upon the bottom ; at others it is moored to the surrounding vegetation. On large sheets of fresh water. The bird breeds on the Norfolk and Suffolk Broads, in Wales, Yorkshire, Shropshire, Cheshire, Lancashire, Surrey, and several other counties. This species has extended its range considerably during recent years, and is now found breeding in several parts of Scotland and Ireland.

*Materials.*—Flags, sedge leaves, reeds, and all kinds of dead water-plants heaped together. The

GREAT CRESTED GREBE'S NEST AND EGGS.

nest has a slight hollow on the top, and does not stand far above the level of the water.

*Eggs.*—Three to five, usually four. White when originally laid, but soon becoming stained and dirtied. Size about 2.2 by 1.45 in.

*Time.*—April, May, and June.

*Remarks.*—Resident but wandering. Note: a harsh, single-syllabled kind of croak. Local and other names: Gaunt Molrooken, Loon, Tippet Grebe, Greater Loon, Cargoose. Gregarious. Covers over eggs on leaving nest, by swift side to side movements of her bill, with nesting materials. Makes several mock nests, supposed to be either for the male, as outlook posts, or for the young ones when hatched. Sits lightly, and dives when the nest is approached.

---

## GREBE, LITTLE. *Also* DABCHICK.

(*Podicipes fluviatilis.*)

Order PYGOPODES; Family PODICIPEDIDÆ (GREBES).

LITTLE GREBE ON NEST.

*Description of Parent Birds.*—Length about ten inches. Bill not very long, straight, and brown. Irides reddish-brown. Crown, back of neck, and the whole of the upper parts, dark rusty-brown. Cheeks, throat, and sides of neck, reddish-brown; breast, belly, and under parts, greyish-white. Legs and toes, dark greenish. The wings are short, tail almost nil, and legs situated far behind.

The female is very similar to the male.

*Situation and Locality.*—Amongst reeds, rushes,

LITTLE GREBE'S NEST (COVERED).

LITTLE GREBE'S NEST (UNCOVERED).

weeds, and long, coarse grass growing on or near the banks of pools, sluggish rivers, lakes, lochs, reservoirs, and canals. The nest is a kind of raft moored amongst stems or built upon submerged branches, sometimes it rests upon the bottom of a shallow pond, in all suitable localities throughout the British Isles.

*Materials.*—A liberal collection of dead, half-rotten, aquatic weeds, thoroughly saturated with water; very shallow at the top.

*Eggs.*—Four to six; as many as seven have upon a few occasions been found. White, and rough-surfaced when first laid, but gradually becoming stained and discoloured by contact with the bird and the decaying weeds upon which they are deposited and are often covered by. Size about 1.45 by 1.0 in.

*Time.*—March, April, May, June, July, and August.

*Remarks.*—Resident, but subject to local movement. Note: alarm, *whit, whit.* Local and other names: Dabchick, Black-chin Grebe, Small *D*ucker, *D*idapper, Dobchick, Loon, *D*ipper (the proper name of an entirely different species). Not a close sitter, but covers over its eggs when voluntarily leaving the nest.

---

## GREENFINCH.

(*Ligurinus chloris.*)

Order PASSERES; Family FRINGILLIDÆ (FINCHES).

*Description of Parent Birds.*—Length about six inches. Bill, short, thick, and flesh-coloured. Irides hazel. Head, neck, back, rump, and upper tail-coverts, yellowish-green, mixed with ashy-grey on the sides of the head and neck, and with greyish-brown

GREENFINCH'S NEST AND EGGS.

on the other parts. The forehead and rump are bright golden-green. Wing-quills dusky, some of them bordered with yellow and others with grey on the outer webs. Tail feathers dusky, those in the middle uniform, the rest bordered with yellow on their exterior webs. Chin, throat, and breast, bright yellowish-green; belly lighter and mixed with ash-grey; vent and under-tail-coverts white, tinged with pale yellow. Legs, toes, and claws light pinkish-brown.

GREENFINCH AND GREAT TIT FEEDING ON SUNFLOWER SEEDS.

The female is somewhat smaller, and her upper parts are greenish-brown, tinged only with yellow on the wing-coverts, rump, and wing and tail quills; but this is of a duller character than that found on the feathers of the male. Under-parts dull greyish-brown, inclining to greenish-yellow on the belly.

*Situation and Locality.*—In thick, whitethorn hedges, gorse bushes, yew-trees, ivy, holly, and other evergreens; in shrubberies, orchards, on commons, and almost anywhere in wooded districts. It is met with in suitable localities throughout the United Kingdom.

*Materials.*—Slender twigs, rootlets, moss, and grass, lined internally with hair and feathers.

*Eggs.*—Four to six, white, pale grey, or white tinged with blue, in ground colour, sparingly spotted with varying shades of brown, from greyish to dark liver-coloured. The spots and markings are generally most numerous at the larger end. Specimens have sometimes been found pure white and unmarked. They are often very difficult to distinguish from the eggs of the Goldfinch. Size about .82 by .56 in. (*See* Plate II.)

*Time.*—April, May, June, July, and sometimes as late even as August.

*Remarks.*—Resident. Notes: call when flying, *yack-yack*, and when perched *shwoing*, according to Bechstein. When disturbed whilst sitting, it utters a heart-softening sort of melancholy *tway* that is enough to fill any young collector with remorse. Local and other names: Green Linnet, Green Chub, Green Grosbeak, Green Bird, Green Lintie. Sits very closely.

---

## GREENSHANK.

(*Totanus canescens.*)

Order Limicolæ; Family Scolopacidæ (SNIPES).

*Description of Parent Birds.*—Length about twelve or thirteen inches. Bill long, slightly curved upwards, and almost black in colour. Irides hazel. Head, sides, and back of neck greyish-white, marked with almost black longitudinal lines. Back and wings (except primaries, which are dull black) greenish-black, each feather being bordered with buffy-white. Tail-quills white, barred in the middle and striped on the outside with ash-brown. Chin, throat, breast, sides, belly, vent, and under tail-coverts white, the throat and sides being slightly streaked with ash-grey. Legs and toes olive-green; claws black.

Female similar to male.

*Situation and Locality.*—On the ground amongst

GREENSHANK'S NEST AND EGGS.

tufts of coarse grass, heather, between dry mounds near lochs and streams in the north and west of Scotland, the Hebrides, and Shetlands.

*Materials.*—A few bits of dead grass, used as a lining to the declivity chosen.

*Eggs.*—Four; pale yellowish-green to warm stone-colour or buff, beautifully blotched or spotted with light purple, grey, and dark brown. Markings most numerous at larger end. Size about 1.95 by 1.35 in. (*See* Plate IX.)

*Time.*—May and June.

*Remarks.*—Migratory, arriving on its breeding-grounds at the end of April or beginning of May, and leaving in July. The bird is said to winter in Ireland. Note: a loud *vir-too', vir-too', vir-too'*, or *chee-weet, chee-weet*, sounding in some respects like that of the Redshank. Local and other names: Cinereous Godwit, Green-legged Horseman, Greater Plover. Sits lightly, and flies straight away from the neighbourhood of the nest, which is very difficult to find. Our photograph was, as ornithologists will understand, not secured without considerable trouble.

---

## GROUSE, BLACK.

(*Tetrao tetrix.*)

Order GALLINÆ; Family TETRAONIDÆ (GROUSE).

*Description of Parent Birds.*—Length about twenty-two inches. Bill short, curved downwards, and black. Irides dark brown. Bare, erectile skin over eyes, bright scarlet. Head, neck, back, wing-coverts, rump, and tail black, richly glossed in parts with blue-black. Wings brownish-black, with a conspicuous white bar across the middle. The tail-feathers are elongated on either side, and form an outward kind of curving hook. Chin, breast, belly, and flanks black; vent, thighs, and legs dark-brown, mixed with white; under tail-coverts white; toes and claws blackish-brown.

The female is shorter by four or five inches, and differs considerably in appearance. Her bill is dark brown. Irides hazel. Plumage red or rusty-brown, barred and freckled with black; the markings are largest on the breast, where the feathers are bordered with greyish-white. The tail is not forked, and the feathers are variegated with rusty-red and black, and tipped with white. Under tail-coverts nearly white. Legs mottled brown; toes and claws brown.

*Situation and Locality.*—On the ground, under tufts of dead bracken, brambles, heather, rushes, and coarse grass. I have seen nests quite exposed in open pasture land, and have known cows tread upon and break their eggs in such situations. On rough broken land containing heather, rushes, ling, gorse, juniper, mixed woods and young plantations. The bird breeds in suitable parts of England, Wales, and Scotland, but not in Ireland. Our illustration is from a photograph taken in the Highlands of Scotland.

*Materials.*—Dry grass, bents, fern or bracken fronds, and other suitable materials at hand, forming a scant lining to the selected hollow.

*Eggs.*—Five to ten; yellowish-white to yellowish-brown, irregularly spotted with smallish red-brown spots. Size about 2.0 by 1.4 in. Distinguished from those of the Capercaillie by their smaller size. (*See* Plate XV.)

*Time.*—April, May, and June.

BLACK GROUSE'S NEST AND EGGS.

*Remarks.*—Resident. Notes: male, a loud cooing, followed by a hissing sound; female's response plaintive. Local and other names: Black Game, Heath Cock, Black Cock, Heath Poult, Grey Hen (female), Brown Hen (female). Sits closely.

---

## GROUSE, RED.

(*Lagopus scoticus.*)

Order GALLINÆ; Family TETRAONIDÆ (GROUSE).

*Description of Parent Birds.*—Length about sixteen inches. Beak short, curved downward, and black. Irides hazel. Above the eye is a scarlet, arched membrane. The dominating colour of the head, neck, back, wing, and tail-coverts is reddish-brown, speckled and barred with black. Wing and tail-quills blackish-brown. Chin and throat rich, dark chestnut-brown, unspotted; breast dark reddish-brown, sometimes almost black; belly, sides, vent, and under tail-coverts light reddish-brown, tipped with white. Legs and toes covered with

RED GROUSE ON NEST.

RED GROUSE'S NEST AND EGGS.

short soft feathers of greyish-white; claws long and horn colour, dark at the base.

The female is somewhat smaller and of a lighter rufous-brown. The membrane over her eye is narrower and less conspicuous. Both male and female are subject to considerable variation in plumage. I have seen some hens a beautiful golden-yellow, and both sexes often showing a considerable amount of white in their plumage.

YOUNG RED GROUSE.

*Situation and Locality.*—A slight hollow or natural depression, generally well hidden by heather or ling, occasionally amongst rushes or long coarse grass on wild moors in Wales, the six northern counties of England, and Derbyshire, Cheshire, Shropshire and Stafford, also in every county of Scotland, excepting perhaps one, and in suitable parts of Ireland.

*Materials.*—A few heather or ling shoots, or bits of bent grass.

*Eggs.*—Five to nine; as many as thirteen to fifteen have been found. Of a dirty white ground-colour, thickly blotched and spotted with umber-brown. Variable in regard to colour and markings,

HEDGE SPARROW.
(*See p.* 376.)

SPOTTED FLYCATCHER.
(*See p.* 117.)

PIED FLYCATCHER.
(*See p.* 114.)

SWALLOW.
(*See p.* 392.)

CUCKOO.
(*See p.* 59.)

PEREGRINE FALCON.
(*See p.* 111.)

CUCKOO.
(*See p.* 59.)

HOBBY
(*See p.* 188.)

NIGHTJAR.
(*See p.* 236.)

MERLIN.
(*See p.* 228.)

NOTE.—*In referring to the eggs the above names should be read fro n left to right.*

but distinguished from those of the Ptarmigan by being less buff in ground-colour and more spotted. Size about 1.75 by 1.25 in. (*See* Plate XV.)

*Time.*—Eggs have been found as early as February and as late as July; but April, May, and June are the principal breeding months.

*Remarks.*—Resident. Notes: female call, *yow, yow, yow,* pronounced with a peculiar nasal catch; crow notes, *cabow, cabow, cabeck, cabeck, beck, beck; cockaway, cockaway;* alarm note of male, *cock, cock, cock.* Local and other names: Gorcock, Moorfowl, Moorcock, Moorgame. A close sitter, resorting to decoy methods when disturbed.

---

## GROUSE, SAND.

(*Syrrhaptes paradoxus.*)

Order PTEROCLETES; Family PTEROCLIDÆ (SAND GROUSE).

Although several great incursions of this species have from time to time taken place, and the bird during its latest, in 1888, bred with us, it has no rightful claim for inclusion at length in this work.

---

## GUILLEMOT, BLACK.

(*Uria grylle.*)

Order PYGOPODES; Family ALCIDÆ (AUKS).

BLACK GUILLEMOT.

*Description of Parent Birds.*—Length about fourteen inches. Bill fairly long, straight, and black. Irides brown. The whole of the plumage is black, with exception of a large patch on the coverts of each wing, which is white. Legs, toes, and webs vermilion-red; claws black.

BLACK GUILLEMOT'S EGGS.

The female is similar to the male in size and coloration.

*Situation and Locality.*—In deep crevices of rocks overhanging the sea; amongst large stones heaped loosely together; and occasionally under or between crags and large fragments of rock near the beach. Principally on the western and northern coasts of Scotland and the islands round about it, in a few suitable places round the coast of Ireland, and to a limited extent in the Isle of Man.

*Materials.*—None; the eggs being laid on the bare rock or ground.

*Eggs.*—Two, white, faintly tinged with green, blue, or creamy-buff, spotted and blotched with ash-grey, reddish or chestnut-brown, and very dark brown. Size about 2.35 by 1.6 in. (*See* Plate XIII.)

*Time.*—May and June.

*Remarks.*—Resident, but a southern wanderer in winter. Note: a plaintive whine. Local and other names: Sea Turtle, Greenland Dove, Dovekie Scraber, Tyste, Puffinet. Gregarious. Sits closely. Keeps to the open sea, except during the breeding season and when driven ashore by stress of weather.

YOUNG BLACK GUILLEMOT.

## GUILLEMOT, COMMON.

(*Uria troile.*)

Order PYGOPODES; Family ALCIDÆ (AUKS).

COMMON GUILLEMOTS.

*Description of Parent Bird.*—Length about eighteen inches. Bill rather long, straight, sharp-pointed (which easily distinguishes the bird from the Razorbill), and black. Irides dusky. Head, neck, back, wings (except ends of secondaries, which are tipped with white), and tail dark mouse-brown. Lower part of throat, breast, and belly white. Legs and feet, which are webbed, brownish black.

The female is rather smaller than the male.

*Situation and Locality.*—On ledges and in hollows of cliffs, on the flat bare summits of rockstacks, in suitable places pretty generally round our coasts. Our full-page illustration shows a great number of these birds sitting on their eggs on the Pinnacles at the Farne Islands. How individuals recognise their own eggs in such a vast crowd one cannot imagine.

*Materials.*—None whatever, the egg being laid on the bare rock.

*Egg.*—One, very large for the size of the bird, and pear-shaped. The eggs of this species present an endless variety of coloration. Sometimes the ground-colour is white, at others cream, yellowish-green, reddish-brown, pea-green-blue, purplish-brown, and every variety of shade between these colours, spotted, blotched, and streaked profusely

COMMON GUILLEMOTS BREEDING ON THE PINNACLES AT THE FARNE ISLANDS.

with black, dusky-brown, greyish-brown, and other tints in great variety. Some specimens are without spots, and I have seen others on Ailsa Craig and elsewhere closely resembling those of the Razorbill, but always more pyriform. Size about 3.25 by 1.95 in. (*See* Plate XII.)

*Time.*—May and June.

*Remarks.*—Resident. Notes (of young): *willock, willock.* Local and other names: Foolish Guillemot, Willock, Tinkershere, Scout, Tarrock, Lavy, Murre, Sea Hen, Marrock. A close sitter.

---

## GULL, BLACK-HEADED.

(*Larus ridibundus.*)

Order GAVIÆ; Family LARIDÆ (GULLS).

BLACK-HEADED GULL.

*Description of Parent Birds.*—Length about sixteen inches. Bill moderately long, nearly straight, and lake-red. Irides hazel. Eyelids crimson. Head and upper part of throat dark brown. Back and sides of neck white. Back and wings (except some of the primaries, which are black at the tips, and on some of the margins with white shafts), uniform lavender-grey. Tail-coverts and quills white. Lower front of neck, breast, and all under-parts, white. Legs and feet lake-red; claws black.

The female is similar to the male.

The above description is of a solitary pair shot

BLACK-HEADED GULL'S NEST AND EGGS.

whilst nesting in June on a northern moorland tarn. The Black-headed Gull is subject to considerable variation in plumage, not only in regard to season but age.

*Situation and Locality.*—On the ground, in a tussock of coarse grass, tuft of rushes, or a slight hollow on the bare ground; in swamps and bogs, at the edges of and on islands in tarns and lakes. In large colonies at a great number of suitable places throughout the British Isles. Three famous places in England are Scoulton Mere in Norfolk, where the bird has nested in thousands for upwards of three hundred years in succession, at Pallinsburn in Northumberland, and at Ravenglass in Cumberland. Although gregarious, I have frequently met with solitary pairs nesting on small mountain tarns.

*Materials.*—Sedges, rushes, tops of reeds, and withered grass; generally in small quantities, sometimes quite absent, and at others in fairly large quantities, much depending upon the site chosen.

*Eggs.*—Two or three; usually the second number, and occasionally four, varying from pale olive-green to light umber-brown in ground-colour, blotched, spotted, and streaked with blackish-brown and dark grey. Size about 2.2 by 1.45 in. They are subject to great variation in regard to size, shape, and colour; but their small size and the presence of the parent birds easily distinguish them. (*See* Plate X.)

*Time.*—April, May, and sometimes as late as June.

*Remarks.*—Resident, but subject to much local movement. Note: a hoarse cackle, resembling a laugh when quickly repeated. Local and other names: Red-legged Gull, Laughing Gull, Peewit

Gull, Blackcap, Sea Crow, Hooded Mew, Brown-headed Gull, Mire Crow, Croker, Pickmire. Sits lightly, and clamours noisily overhead when disturbed. Gregarious, as a rule.

BLACK-HEADED GULLS AT HOME.

---

## GULL, COMMON.

(*Larus canus.*)

Order GAVIÆ ; Family LARIDÆ (GULLS).

COMMON GULL.

*Description of Parent Birds.*—Length about eighteen inches. Bill rather short, slightly curved downward at the tip, and yellow in colour. Irides orange-brown. Head and neck snowy-white. Back and wings French grey ; tips of wings black, spotted with white, on account of some of the primaries having white ends. Tail-coverts and quills snowy-white. Chin, throat, breast, belly, and vent snowy-white. Legs, toes, and webs greenish-yellow.

The female is similar in plumage, but slightly smaller in size.

*Situation and Locality.*—On the ground amongst heather and coarse grass; on ledges and in crevices of rock round the coast of Scotland; on islands; in inland lochs and tarns; also in suitable places in Ireland, but now nowhere in either England or

YOUNG COMMON GULL HIDING

Wales. Our illustrations were procured on the west coast of Scotland.

*Materials.*—Heather, dry seaweed, and dead grass. It may be observed that a somewhat large nest is built as a rule.

*Eggs.*—Two to four; generally three, buffish-brown or dark olive-brown in ground-colour, spotted, blotched, and streaked with grey, dark brown, and black, irregularly distributed. Size about 2.25 by 1.65 in. The smallness of the spots and the size of the eggs enable the student to easily identify them. (*See* Plate XI.)

*Time.*—May and June.

*Remarks.*—Resident, but subject to local

COMMON GULL'S NEST AND EGGS.

movement. Notes: a kind of squeal. Local and other names: Winter Mew, Sea Mew, Sea Mall or Maw, Sea Gull, Sea Cob, Blue Maa. Gregarious. A light sitter, and clamorous when disturbed.

---

## GULL, GREAT BLACK-BACKED.

(*Larus marinus.*)

Order GAVIÆ; Family LARIDÆ (GULLS).

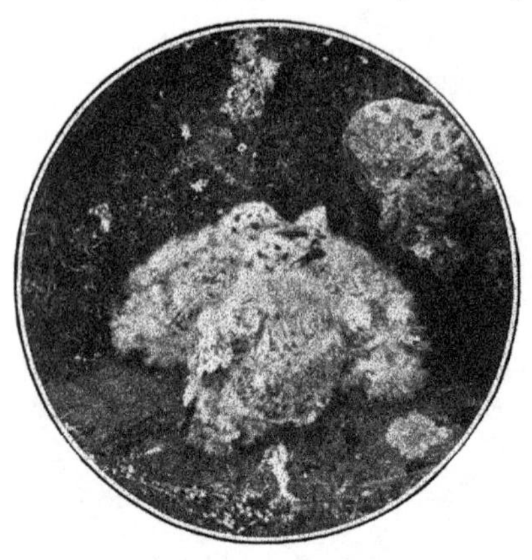

YOUNG GREAT BLACK-BACKED GULLS.

*Description of Parent Birds.*—Length about thirty inches. Bill of medium length, large and powerful; pale yellow, excepting a portion of the under mandible, which is orange; upper mandible turned down at the tip. Irides straw-yellow. Head and neck all round snowy-white. Back and wings black, with exception of the tips of the quills, which are white. Upper tail-coverts and tail-quills white. Breast and all under-parts pure white. Legs, toes, and webs pale flesh-colour.

The female is similar, but somewhat smaller.

*Situation and Locality.*—On the ledges of maritime cliffs, on the tops of rock stacks and islets in the sea and fresh-water lakes; also on the ground, in marshes and moors; on the coasts of Dorset, Cornwall, Scilly, and Lundy; on the Welsh coast, but most abundant on the western and northern shores of Scotland and the islands lying round about; also in Ireland.

GREAT BLACK-BACKED GULL'S NEST AND EGGS.

*Materials.*—Seaweed, heather, wool, and dry grass in variable quantities. Sometimes they are very abundant and at others almost entirely absent.

*Eggs.*—Two or three, generally the latter number. Yellowish-brown or stone-colour to light olive-brown, blotched with slate-grey and dark brown. The spots are not very large, and generally distributed over the surface of the egg. Size about 3.1 by 2.1 in. The large size of the eggs and the small spots are distinguishing characteristics. (*See* Plate X.)

*Time.*—May and June.

*Remarks.*—Resident, but wandering during the non-breeding months. Note: a harsh croak or laugh. Local and other names: Cob, Blackback, Great Black Salmon Gull and White Gull. Gregarious in some parts and solitary in others. Not a close sitter, but demonstrative when intruded upon. The bird figured in our tailpiece in the act of alighting was exceedingly shy and suspicious, although she could see nothing but the lens of the camera at a considerable distance from her.

GREAT BLACK-BACKED GULL ALIGHTING.

## GULL, HERRING.

(*Larus argentatus.*)

Order GAVIÆ ; Family LARIDÆ (GULLS).

HERRING GULL.

*Description of Parent Birds.*—Length about twenty-four inches. Bill of medium length, hooked at the tip, and yellow, with an orange spot on the lower mandible. Irides pale yellow. Head and neck white. Back and part of wings light grey, which distinguishes the bird from the Lesser Black-backed Gull; quills blackish, tipped with white. Breast, belly, vent, upper tail-coverts and tail-quills pure white. Legs and feet flesh-colour. Variable with age.

The female is often much smaller than the male, but is similar in the coloration of her plumage.

*Situation and Locality.*—Ledges of sea cliffs, low rocky islands, sometimes in marshes, such as Foulshaw Moss in Westmorland. At the Farne Islands a few pairs only of Herring Gulls breed amongst the great crowd of Lesser Black-backed Gulls, but on islands in Hebridean lochs the case is absolutely reversed, the former far outnumbering the latter.

*Materials.*—Seaweed and turf, lined with grass, sometimes in liberal quantities, at others very scant, or absent altogether. The grass used appears to have often been obtained quite green. In the summer of 1905 I built a hiding structure of stones

HERRING GULL'S NEST AND EGGS.

for my camera and myself on a small island in a Hebridean loch and covered it over with heather. Two or three days afterwards I found that several Herring Gulls had stolen my thatch and made their nests with it.

*Eggs.*—Two or three, varying in ground-colour from olive-green to buffish-brown, spotted and blotched with dark brown and grey. Variable, and bearing a very close resemblance to the eggs of the Lesser Black-backed Gull, but said to run larger by some authorities, and to be more spotted and less blotched. The fishermen reckon to distinguish them from those of the above-mentioned species by the paler ground-colour of the shell; but the only sure method of identification I have found is to watch the parent bird on to her nest. Average size about 2.85 by 2.0 in. (*See* Plate X.)

*Time.*—May and June.

*Remarks.*—Resident. Call-notes, *hau-hau-hau*; alarm, *ky-eok*. Local or other name: none. Sits lightly, and watches the intruder closely from a distance of eighty or a hundred yards. A terrible egg-stealer.

HERRING GULLS AT HOME.

## GULL, KITTIWAKE. *See* KITTIWAKE.

*See* KITTIWAKE.

---

## GULL, LESSER BLACK-BACKED.

(*Larus fuscus.*)

Order GAVIÆ ; Family LARIDÆ (GULLS).

LESSER BLACK-BACKED GULLS.

*Description of Parent Birds.*—Length about twenty-three inches. Beak of medium length, nearly straight, and yellow, with the exception of an orange spot on the under mandible. Irides straw colour. Head and neck all round pure white. Back and wings dark slate-grey, some of the quills being slightly tipped with white. Upper tail-coverts and tail-quills white. Breast, belly, vent, and under tail-coverts pure white. Legs and feet yellow, whereas those of the Great Black-backed Gull are flesh colour.

The female is said to be a little smaller, and the feathers of the back and wings to vary much in tint with age and locality.

*Situation and Locality.*—On the ground, in hollows scooped out of the soft turf, on grass growing in nooks and on ledges of rock, on bare rocks, and on masses of dry seaweed. Our illustrations are from photographs taken on the Farne Islands, where a large colony breeds. On low rocky islands, ledges of cliffs, on islands in inland lakes, and in moss-bogs. At nearly all suitable places round our coasts, except on the eastern and southern shores of England.

LESSER BLACK-BACKED GULL'S NEST AND EGGS.

*Materials.*—Seaweed, often in large quantities; grass, which appears to have been collected quite green; sometimes no materials whatever, the eggs being laid on the grass in a hollow.

*Eggs.*—Two to four, generally three. Very variable, from light drab to dark olive-brown; sometimes pale bluish-green, spotted, blotched, and streaked with ash-grey, pale brown, and dark liver-brown. Size about 2.6 by 1.85 in. (*See* Plate XI.)

*Time.*—May and June.

*Remarks.*—Resident, but subject to much local movement. Notes: call, *ha, ha, ha*, or *an, an, an*; note of anger, *kyeok*. Local and other names: Yellow-legged Gull, Less Black-backer Gull. Not a very close sitter, but noisy and clamorous when disturbed. Gregarious. Under protective measures this species grows quite bold, proof of which is given in such breeding stations as the Farne Islands.

LESSER BLACK-BACKED GULLS FLYING OVER THEIR NESTS

HARRIER, ASH-COLOURED. *See* HARRIER, MONTAGU'S.

---

## HARRIER, HEN.

(*Circus cyaneus.*)

Order ACCIPITRES; Family FALCONIDÆ (FALCONS).

*Description of Parent Birds.*—Length about eighteen inches. Beak short, much curved, bluish-black, and surrounded at the base with black, bristly feathers. Bare skin immediately round base of beak, and irides yellow. Head, neck, back, wings, and upper side of tail bluish- or ash-grey, except the wing-primaries, which are almost black. Some specimens have a mottled, rusty-brown spot on the nape. Chin, throat, breast, and belly bluish-grey, much lighter on the latter parts. Thighs, vent, and under tail-coverts white. Under-side of tail-quills very light grey, faintly barred with a darker tinge. Legs and toes yellow; claws black.

The female measures about four inches longer; her bill is nearly black, and the bare skin round the base tinged with green. Irides reddish-brown. Crown and back of neck dark brown; round the face is a kind of ruff, the feathers of which are a mixture of brown and white. Back and wings umber-brown, except some of the coverts, which are edged with rufous, and the primaries, which are of a dusky colour. The tail-quills are dark brown tipped with rusty-red; the centre ones uniform in colour, and those on the sides barred with lighter rusty-brown. Throat and all the under-parts reddish-buff, with a darker patch in the centre of each feather. Tail-feathers underneath barred with brownish-black and grey.

*Situation and Locality.*—On the ground, amongst tall heather, furze, and other bushes; on moors, commons, fens, and on wild, lonely mountain-sides. Its destructive habits amongst game birds have made the gamekeeper an especial enemy, and he has waged incessant war upon it for so long that it is now almost exterminated in England. I have seen its nest in Surrey during 1907, and am pleased to say that the young ones got away safely. It is said to breed in Cornwall, Devon, Somerset, and one or two other western counties, Wales, and the north of England occasionally. Its nest occurs most frequently in the Hebrides, Orkneys, and Highlands of Scotland; also in suitable parts of Ireland.

*Materials.*—Small sticks, sprigs of heather, and coarse grass; in sparing quantities where the nest is placed in a dry situation; but when a low, damp place is chosen, sticks, reeds, sedge, and flags are used in liberal quantities.

*Eggs.*—Four or five, occasionally six. White, faintly tinged with blue or bluish-green; on rare occasions slightly marked with light rusty-red or yellowish-brown. They vary in size, and closely resemble those of the Marsh and Montagu's Harriers. Size about 1.75 by 1.45 in.

*Time.*—May and June.

*Remarks.*—Formerly resident, now probably only migratory. It arrives in April or May, and departs in September and October. Notes: tremulous and Kestrel-like. Local and other names: Male, Dove Hawk, Blue Hawk, or Miller; female, Ringtail; and in the Hebrides a Gaelic name signifying

HEN HARRIER'S NEST AND EGGS

Mouse Hawk. The sexual difference in plumage was the cause of the birds being believed at one time to represent different species. Not a close sitter.

---

## HARRIER, MARSH.

(*Circus æruginosus.*)

Order Accipitres; Family Falconidæ (FALCONS).

*Description of Parent Birds.*—Length about twenty-one inches. Beak short, curved, and bluish-black. Bare skin round the base of the beak, and irides yellow. Crown, sides of head, and nape pale rusty yellowish-white, streaked with darkish-brown. Back dark brown tinged with red, the feathers being bordered with a lighter shade. Wing-coverts and tertials varying, according to age, from dark reddish-brown to ash-grey; secondaries ash-grey; primaries varying from brownish-black to slate-grey. Tail ash-grey. Chin and throat almost white; breast and under-parts reddish-brown, streaked with dark brown. Legs and toes yellow; claws black.

The female is larger, and slightly duller in her plumage. Both are subject to variation in colour. according to age.

*Situation and Locality.*—On the ground, amongst sedges, reeds, ferns, and under furze and other small bushes; rarely in trees. On low, marshy, reed- and water-covered land; also unfrequented moors. Professor Newton, in the latest edition of Yarrell, issued 1874, says that "the bird breeds regularly in Devonshire, Norfolk, and Aberdeenshire"; and Mr. Dixon, in his "Nests and Eggs of British Birds," issued just twenty years after, says that Norfolk is the only county in Great Britain in

MARSH HARRIER'S NEST YOUNG AND EGGS.

which the bird regularly attempts to breed. This is one among many of the facts which serve to illustrate the rapidity with which our rarer birds are being banished. Our picture of a nest containing a newly hatched young one and eggs was secured by my brother in Holland. And the tailpiece to this article represents the place where a few years ago probably the last attempt was made by a pair of these rare birds to breed in Norfolk. The nest was built, but before the unfortunate hen had a chance of laying in it she was ruthlessly slain.

*Materials.*—Sticks, twigs, rushes, and reeds in rather large quantities, lined with dead grass.

*Eggs.*—Three to five or six. White, sometimes slightly tinged with pale bluish-green or milk-blue, and upon rare occasions marked with a few spots of rusty-red. Size about 1.95 by 1.55 in.

*Time.*—May.

*Remarks.*—Resident, but wandering. Notes: male, *koi* or *kai*; female, *pitz pitz, peep peep.* Local and other names: Duck Hawk, White-headed Harpy, Moor Harrier, Moor Buzzard, Puttock, Marsh Hawk, Bald Buzzard. Sits lightly.

HOME OF THE MARSH HARRIER.

## HARRIER, MONTAGU'S. *Also* Ash-Coloured Harrier.

(*Circus cineraceus.*)

Order Accipitres ; Family Falconidæ (FALCONS).

*Description of Parent Birds.*—Length about seventeen inches. Beak short, upper mandible much curved and nearly black. Skin round base of beak bare, and greenish-yellow. Irides bright yellow. Head, neck, back, and wing-coverts bluish-grey. Primaries nearly black; secondaries marked by three bars. Tail-quills, on the sides, white, barred with bright rust colour; centre feathers bluish-grey. Chin and throat brownish-grey; breast, belly, and under-parts white, streaked with bright rust colour. Legs and toes yellow; claws black. The wings are very long and narrow.

The female is about nineteen inches long. Beak black; bare skin at base, dull yellow. Irides hazel. Crown and back of head reddish-brown, with spots of a darker tinge. Over and under the eye is a streak of grey. Back and wings dark umber-brown; rump and upper tail-coverts orange-brown and white. Side feathers of tail barred with brown of two shades; breast and all under-parts light reddish-brown; claws black. Both sexes of this bird vary considerably, according to age and individual.

*Situation and Locality.*—On the ground, amongst heather, ferns, long grass or rushes, furze, and low brushwood; on moors and heaths in Norfolk, Kent, Pembrokeshire, Dorsetshire, Hampshire, Devonshire, and Somerset. Very rare, and on a fair way to total extinction, so far as the British Isles are concerned.

NEST AND EGGS OF MONTAGU'S HARRIER.

The nest figured in our illustration was situated not very far from the unfortunate Marsh Harrier's attempt at housekeeping mentioned in the article dealing with that species, but was not built in such deep sedge grass. It was placed flat upon the wet marsh ground, and had a sort of little courtyard in front of it, where all the vegetation had been cleared away or beaten down. The diameter of the actual structure was about eight and a half inches, and its materials consisted of rushes, sedge, ragwort stems, and a few bits of dead grass. The Rev. M. C. H. Bird considered it a large nest for two eggs, as the birds add materials, like many other species, as they lay their eggs.

*Materials.*—Twigs, heather-stalks, straws, and dry grass, sometimes wool, scantily and loosely lining the slight hollow chosen for the reception of the eggs.

*Eggs.*—Four to six. Very pale bluish-white, said to be sometimes marked with a few spots of pale reddish-brown. Average size about 1.65 by 1.4 in.

*Time.*—May.

*Remarks.*—Migratory, arriving in April and leaving in October. Notes: something like those of the Kestrel, but feebler and more querulous, according to Mr. Saunders. This species only enjoys one local or alternative name so far as I can gather, and that is the Ash-coloured Harrier. Sits lightly. The bird's destructive habit of feeding upon eggs supplies it with an uncompromising enemy in the gamekeeper.

## HAWK, SPARROW. *See* Sparrow-Hawk.

---

## HAWFINCH.

(*Coccothraustes vulgaris.*)

Order Passeres ; Family Fringillidæ (FINCHES).

*Description of Parent Birds.*—Length about seven inches. Bill of medium length, nearly conical, very thick at the base, and of a dusky-blue colour. Irides grey. Crown and sides of head dull yellowish- or orange-brown ; back and sides of neck ash colour. Back, smaller wing-coverts, and scapulars chestnut-brown. Some of the middle wing-coverts are white ; wing-quills black, glossed with blue ; some of them are of curious appearance, suggesting that they have been clipped at the tips so as to form battleaxes or billhooks. Rump and upper tail-coverts light orange-brown ; tail-quills black, the outer ones being tipped and to some extent edged with white ; middle greyish-brown, tipped with white. The feathers round the base of the beak, eyes, and on the throat are black ; breast and belly pale rust colour ; vent and under tail-coverts dull white. Legs, toes, and claws pale brown.

The female is less brilliant, and her colours are more mixed.

*Situation and Locality.*—In old lichen-covered hawthorn bushes ; on the horizontal branches of oaks, heads of pollards ; in holly bushes, firs, fruit and other trees, at varying heights, in gardens, orchards, timbered commons, and plantations, pretty generally, though not commonly, in all the counties of England. I have met with it

HAWFINCH'S NEST AND EGGS.

on Surrey commons, but never saw its nest in any part of Yorkshire, where it undoubtedly breeds. A friend of mine has found its nest in the neighbourhood of Bedale. It is very common in some parts of Kent, where as many as four nests in one orchard have been found at the same time.

*Materials.*—Twigs, fibrous roots, and grass, mixed with lichens, and lined internally with fine fibrous roots, grass, and hair; somewhat loosely constructed.

*Eggs.*—Four to six, pale olive-green, varying to pale reddish-brown, or greenish-grey, spotted with blackish-brown, and irregularly streaked with dusky-grey. Size about .95 by .75 in. (*See* Plate II.)

*Time.*—May, although nests may sometimes be found containing young ones as late as the end of August.

*Remarks.*—Resident, although its numbers are increased in winter by Continental arrivals. Notes: call, rendered by Bechstein as an unpleasant *itszip*, uttered incessantly; song, a light jingle, with some clearer, shrill, and harsh notes like *irrr*. Local and other names: Grosbeak Haw, Grosbeak, Common Grosbeak, Black-throated Grosbeak. Sits closely. This species would be much commoner in the British Isles were it not for the fact that it has a fatal weakness for green peas and is destroyed in considerable numbers by gardeners in defence of their property.

---

HEDGE SPARROW. *See* SPARROW, HEDGE.

OSPREY.

(*See p.* 242.)

KESTREL.

(*See p.* 199.)

GOLDEN EAGLE.

(*See p.* 106.)

NOTE.—*In referring to the eggs the above nanes*

## HERON, COMMON.

(*Ardea cinerea.*)

Order Herodiones ; Family Ardeidæ (HERONS).

HERON.

*Description of Parent Birds.*—Length about thirty-six inches. Beak long, straight, strong, pointed, and dusky in colour, except at the base of the under mandible, where it is yellowish. Irides yellow. Forehead, crown, and cheeks, greyish-white. On the back part of the head the feathers are elongated into a kind of plume, and are bluish-black or dark slaty-blue in colour. Upper surface of body and wing-coverts bluish-grey; wing-primaries black ; tail-quills cinereous. Neck white, adorned with large longitudinal elongated spots of dark bluish-grey in front. On the lower part of the neck the feathers are elongated, and hang loosely down. Breast, belly, thighs, and under-parts greyish-white, streaked with black. Legs and toes dirty yellowish-green ; claws black.

In the female the plumes are shorter, and her colour duller and less distinctive.

*Situation and Locality.*—On the tops of high trees, ledges of cliffs, and in some places even on the ground. The bird has been known to breed in at least forty-one counties of England and Wales, and does so in various suitable parts of Scotland and Ireland.

*Materials.*—A liberal collection of sticks and

twigs, lined with turf, moss, fibrous roots, and sometimes wool or rags, according to some authorities.

*Eggs.*—Three to five, pale blue with a tinge of green. Size about 2.5 by 1.7 in. (*See* Plate VIII.)

*Time.*—January, February, March, and April.

HERON'S NEST.

The two first months only in exceptionally fine, open seasons. I have seen eggs in nests during May and the early part of June in the island of Skye and other parts of the Highlands.

*Remarks.*—Resident. Notes: harsh, short, and guttural. Some naturalists describe the alarm note as *frank, frank, cronk,* but it sounds to me like *garowk, garowk, garowk.* Local and other names: Hearinsew, Hern, Heronshaw, and (in Ireland) Cranes. Gregarious, as many as eighty nests having been known in a single tree. The birds return to the same place (called a Heronry) year after year. Not a very close sitter, as a rule. However, I have known individual birds sit quite still whilst the trunk of a tree in which their nests were situated has been violently struck with a stick.

YOUNG HERONS IN NEST.

---

## HOBBY.

(*Falco subbuteo.*)

Order ACCIPITRES; Family FALCONIDÆ (FALCONS).

*Description of Parent Birds.*—Length about twelve inches. Beak short, much curved, and bluish horn colour. Bare skin round the base of the beak of a greenish-yellow colour. Irides dark brown. Crown, nape, back, and wings greyish-black, the feathers being edged with buffish-white. Wing-quills almost black, bordered with light grey. Tail-quills greyish-black, barred and tipped with a lighter tint, except the two middle feathers, which are uniform greyish-black. Chin and sides of neck white; cheeks black; breast and belly yellowish-white, streaked broadly with brownish-black; thighs, vent, and lower tail-coverts rusty-red. Legs and toes yellow; claws black.

The female resembles the male, but is larger, and the spots on her breast are more conspicuous. In young birds of both sexes the plumage on the upper surface of the body is tinged with red, but this gradually gives place to bluish-grey with age.

*Situation and Locality.*—In high trees in woods and forests. It is almost as local as the Nightingale, and has not been reported as nesting farther north than Yorkshire, except on one occasion in Scotland. I have seen it breeding in Wiltshire, and it also breeds in Surrey. It is a rare nesting species, but returns to a favourite haunt year after year.

*Materials.*—Some high authorities say that it does not build a nest of any kind, but simply adopts the old one of a Carrion Crow, Magpie, Woodpigeon, or that of some other Hawk; whilst others say that it builds a nest of sticks, moss, and hair. From my own limited observation, and inquiries made of those who have had opportunities of forming an opinion, I think the former are right.

*Eggs.*—Two or three, very rarely four, short and oval in form. Yellowish, dingy or bluish-white in ground-colour, much suffused, mottled, and spotted with reddish- and yellowish-brown. The eggs are similar in coloration and variety to those of the Merlin and Kestrel, and, as the nesting situations are similar to those sometimes adopted by the latter bird, nothing short of a sight of the parents can settle the point with certainty. Size about 1.7 by 1.35 in. (*See* Plate V.)

HOBBY'S NEST.

*Time.*—May and June.

*Remarks.*—Migratory, arriving in April and leaving in October. Notes: alarm, a shrill chattering, not unlike that of an angry Merlin. Local or other name: none. Sits lightly, according to Mr. Dixon, and fairly closely, according to Mr. Seebohm. Both are in a measure right, for the bird which owned the nest figured in our illustration sat closely when the friend who showed it to me first discovered it, but during the day I spent at the place she sat very lightly. Is very demonstrative when intruded upon whilst the young are in the nest. On one occasion I saw a brood of chicks of this species leave the nest and the chattering of the adult birds as they circled above the little wood in which they had reared their family was incessant.

---

## HOOPOE.

(*Upupa epops.*)

Order PICARIÆ; Family UPUPIDÆ (HOOPOES).

This bird has been known in past times to breed in several parts of England, but its persecution—which dates from King Solomon's time, when tradition says that its handsome crest was made of gold—has so increased, that a detailed description of the bird, its habits, nest, eggs, etc., seems unnecessary in a work of this character. I fear a breeding pair of birds is never again likely to escape the lynx-eyed gunner in this country.

---

HOUSE SPARROW. *See* SPARROW, COMMON.

## JACKDAW. *Also* DAW.

(*Corvus monedula.*)

Order PASSERES ; Family CORVIDÆ (CROWS).

YOUNG JACKDAW.

*Description of Parent Birds.*—Length about fourteen inches. Beak of medium length, strong, nearly straight, and black. Irides greyish-white. Crown black with a purple sheen ; nape and back of neck leaden-grey. Back, wings, upper tail-coverts, and tail black, glossed with blue, violet, and green. All the under-parts are dusky-black. Legs, toes, and claws black.

The female is a trifle smaller than the male, and the grey on the back of her neck is less pronounced.

*Situation and Locality.*—Holes in cliffs, church steeples, towers, old ruins, barns, in chimneys, and hollow trees, pretty generally throughout the British Isles. Our illustration is of a nest in the ventilation hole of a stone barn. It was slightly drawn forward, and light reflected upon it with a looking-glass, in order to take the photograph. The largest colony I have ever met with is near Armathwaite Castle, in Cumberland.

*Materials.*—Sticks, straw, moss, feathers, wool, down, and all sorts of odds and ends the bird can pick up near at hand. In some situations no sticks or twigs are used, and I have examined nests made entirely of rushes from beginning to end.

*Eggs.*—Three to six, usually five. Pale greenish-blue or bluish-white, spotted, speckled, and blotched

JACKDAW'S NEST AND EGGS.

with dark olive-brown and ash-grey. The markings vary in their distribution, being sometimes evenly distributed and at others collected round the larger end. The ground-colour and markings are also subject to considerable variation. Size about 1.45 by 1.0 in. (*See* Plate I.)

*Time.*—May and June.

*Remarks.*—Resident. Notes: *kae*, or *caw*, and *jack*. Local and other names: Daw, Kae, Jack. Gregarious, and a close sitter.

JACKDAW.

---

## JAY.

(*Garrulus glandarius.*)

Order PASSERES; Family CORVIDÆ (CROWS).

*Description of Parent Birds.*—Length about fourteen inches. Beak rather short, nearly straight, strong, and dusky. Irides white, slightly tinged with blue. Crown greyish-white, spotted and streaked with black and purplish-buff; the feathers form a crest, which the bird can elevate or depress at pleasure. Nape and sides of neck, back, and scapulars purplish-buff. Wing-coverts composed of alternate bars of pale blue, sky-blue, and black. Greater wing-quills black, with greyish-white edges;

secondaries deep black, marked with a white patch on the upper half; rump white; tail dusky. From the gape, backward and downward, runs a moustache-like black dash; throat dirty white; breast pale purplish-buff; belly, vent, and under tail-coverts nearly white. Legs, toes, and claws brown.

YOUNG JAYS IN NEST.

The female is very similar to the male in appearance.

*Situation and Locality.*—In a tall thick bush, hedgerow, or young tree; sometimes in evergreens, such as the yew and holly; in woods and plantations with a thick undergrowth. In suitable localities throughout the British Isles.

*Materials.*—Sticks, small twigs, mud, fibrous roots, and grass. Well built, as a rule, and somewhat like a large Blackbird's nest.

*Eggs.*—Five to seven. Ground-colour dusky-green, tinged with blue, thickly spotted and freckled with light olive-brown. The markings are generally uniformly distributed, but are sometimes confluent at the larger end, where there are occasionally several irregular blackish-brown lines. Size about 1.25 by .9 in. (*See* Plate I.)

JAY AT PHEASANTS' DRINKING PLACE.

JAY'S NEST AND EGGS.

*Time.*—April, May, and June.

*Remarks.*—Resident. Note: a harsh, rapidly delivered kind of chatter, sounding like *rake, rake.* Local and other names: Jay Piet, Jaypie. Not a very close sitter.

---

## KESTREL.

(*Falco tinnunculus.*)

Order Accipitres; Family Falconidæ (FALCONS).

*Description of Parent Birds.*—Length about thirteen inches. Bill short, much curved, and lead-coloured. Bare skin round the base of the beak, yellow. Irides dark brown. Head and nape of the neck ash-grey, under the eye is a dusky streak. Back, scapulars, and wing-coverts brownish-fawn colour, spotted with black; wing-quills black, edged with grey; tail-feathers ash-grey, with a broad black bar near the end, which is tipped with white; under-parts light rust colour, spotted and streaked with black; thighs, vent, and under tail-coverts unspotted. Legs and toes yellow; claws black.

The female is about two inches longer. Her head and tail are reddish-brown, also the back, which is duller than that of the male. On the head are some dark streaks, and the back is barred with bluish-black. The tail is very evenly and prettily barred with black from the base to very near the end, where the bars become broader. Under-parts are fainter than in the case of the male.

*Situation and Locality.*—On ledges and in crevices of sea cliffs and inland crags and precipices; holes in trees, towers, old ruins, church steeples, and even dove-cotes have been utilised; also in deserted

KESTREL'S EGGS IN A RAVEN'S OLD NEST.

nests of Ravens, Crows, Magpies, and Sparrow Hawks throughout the United Kingdom. The "nest" represented with young was situated on the stump of a tree growing horizontally, as near as possible, from the crevice of a Highland precipice.

*Materials.*—Generally none at all, a cavity being scratched in the soft earth and leaves in a crevice or nook, which soon becomes plentifully

YOUNG KESTRELS.

besprinkled with castings. Sticks, grass, and wool are said to be sometimes used, but I have never met with either.

*Eggs.*—Four to seven, generally five or six; dirty creamy-white in ground-colour, thickly blotched and clouded with reddish-brown. Very variable. In some the ground-colour is light brown, darkening towards the larger end, blotched and spotted with a darker shade. The colour in all varieties is, as a rule, most abundant at the larger end. Size about 1.55 by 1.25 in. Indistinguishable from the eggs of the Hobby, and a sight of parent bird only can settle identity. (*See* Plate VI.)

*Time.*—April and May.

*Remarks.*—Migratory, although a good many specimens remain through the winter in the southern and midland parts of England. Note: a chattering kind of scream. Local and other names: Windhover, Staengall, Stannel Hawk, Stannel Hoverhawk, Stonegall, Creshawk, Standgale. Not a very close sitter, according to my experience, although some authorities say it is.

---

## KINGFISHER.

(*Alcedo ispida.*)

Order PICARIÆ; Family ALCEDINIDÆ (KINGFISHERS).

YOUNG KINGFISHER.

*Description of Parent Birds.*—Length about seven inches. Bill long, strong, straight, and black except at the base of the under mandible, where it is orange. Irides hazel. Crown, nape, back, wings, rump, upper tail-coverts, and tail dark, greenish-blue; the head and neck are barred with brilliant azure blue. The wing-coverts are spotted with the same colour, which is prominent on the middle of the back, rump, and upper tail-coverts. Wing-quills dull greenish-black, greenish-blue on the outer webs, and reddish-brown on the outside edges of the inner, except at the tips, which are dull black. From the base of the upper mandible to the eye, and thence to the ear-coverts, chestnut. Chin and throat dirty white, slightly tinged with rust colour. Breast, belly, sides, vent, and under tail-coverts beautiful chestnut; duller on the last two mentioned parts. Legs and toes pink; claws, brownish-black.

The female has a shorter beak, and is slightly duller in her plumage.

*Situation and Locality.*—A hole in river or other bank, generally well hidden by some overhanging piece of earth, undermined by the action of the water; occasionally in the side of a sandpit, or, more rarely still, in a hole in a wall. The hole is from one to three or four feet in length, sloping upwards, and ends in a rounded chamber. The bird, in some instances, excavates it, and when such is the case the hole is said to be oval, with its longest diameter vertical, and in others adopts the old nest of a Sand Martin, or even a rat's hole. Common in most suitable parts of the United Kingdom in spite of persecution. The bird is said to be most abundant in the neighbourhood of Oxford, and absent from the most northern parts of Scotland.

*Materials.*—Fish bones in variable quantities, and the dry mould of the hole.

*Eggs.*—Five to eight, sometimes as many as ten. Of a beautiful pink colour before being blown, on account of the yolk showing through, but snowy-white and glossy afterwards. Size about .9 by .75 in.

*Time.*—February, March, April, May, June, and July.

*Remarks.*—Resident. Note: a piercing metallic whistle, sounding like *te-et,* which never fails to

KINGFISHER'S NESTING HOLE.

attract the naturalist's attention. Local or other name: Halcyon. Sits closely, and generally betrays the whereabouts of its nest by the white droppings near the entrance to the hole.

KINGFISHER WAITING FOR PREY

---

## KITE.

(*Milvus ictinus.*)

Order ACCIPITRES; Family FALCONIDÆ (FALCONS).

KITE'S NEST.

*Description of Parent Birds.*—Length about twenty-six inches. Beak shortish, hooked at the tip, strong, and horn coloured. Bare skin round the base of the beak, and irides yellow. Head and neck light grey, streaked with cinereous brown; back and wing-coverts dusky, bordered with rusty-red. Wing-quills dusky-black, some of the inner ones being edged with white on the interior webs. Upper tail-coverts rusty-red; quills rusty-brown, barred on the inner webs with dusky-brown. The tail is much forked. Under-parts rusty-brown, whitish on the chin,

YOUNG KITES IN NEST.

throat, and under tail-coverts, and streaked with dusky-brown, except on the last-named part. Legs and toes yellow; claws black.

The female is somewhat larger, and is said to be greyer about the head and redder beneath the body. However, some ornithologists say that she is less red than the male.

*Situation and Locality.*—In the forked branch of a tree, or on several branches close to the trunk, at varying heights, in certain parts of Wales, which must remain nameless on account of the shameful persecution to which this struggling species is subjected by men who unfortunately attach more importance to a few faded eggshells in a cabinet than they do to seeing this noble bird gliding along its native mountain sides on outstretched wings.

*Materials.*—Sticks and twigs are used liberally for the outsides, and the foundation is lined with moss, wool, grass, and any rubbish the bird can pick up, such as bits of paper and rags.

*Eggs.*—Two to four, generally three. Greyish dirty white, spotted, blotched, and streaked with dull red and brownish-yellow, with underlying markings of greyish-lilac. The markings are generally most numerous at the larger end. Subject to considerable variation. Size about 2.25 by 1.75 in. (*See* Plate VII.)

*Time.*—May.

*Remarks.*—Resident. Note: a shrill shriek, known in some localities as a "*whew.*" Local and other names: Glead, Fork-tailed Kite, Fork-tailed Glead, Gled or Greedy Gled, Puttock, Crotchet-tailed Puttock, Glade. Sits pretty closely, and it is said will defend its nest when in danger of having it robbed; but my experience whilst photographing nests containing young leads me to doubt the accuracy of this statement.

ANCIENT HOME OF KITE.

---

## KITTIWAKE. *Also* KITTIWAKE GULL.

(*Rissa tridactyla.*)

Order GAVIÆ; Family LARIDÆ (GULLS).

KITTIWAKES.

*Description of Parent Birds.*—Length about fifteen and a half inches. Bill of medium length, slightly curved downward, and greenish-yellow in colour. Irides dusky-brown. Head and neck white; back and wings pale grey, the longest quills of the latter tipped with black. Tail-coverts and quills white. Chin, throat, breast, belly, vent, and under tail-coverts snowy white. Legs, toes, and membranes dusky. The female is similar to the male.

KITTIWAKES ON NESTS AT THE FARNE ISLANDS.

*Situation and Locality.*—In crevices and on ledges of rock, at various heights above the sea. It will be noticed in the full-page picture that some of the ledges are so small in area that the birds have to sit in peculiarly uncomfortable positions in order to cover their eggs. Colonies inhabit a great number of suitable breeding-places round our coasts.

"AMIDST THE FLASHING AND FEATHERY FOAM."

*Materials.*—Heath, dry seaweed, and dead grass, somewhat carelessly arranged.

*Eggs.*—Two to four. Some authorities say that two is the most general number; others three, an opinion which experience leads me to favour. Ground-colour varies from stone-yellow to buffish-brown, sometimes shaded with blue, blotched and spotted thickly with ash-grey, light brown, reddish- or umber-brown. Very variable, both in ground-colour and markings. Size about 2.15 by 1.6 in. The size of the eggs, their large markings, and the situation of the nest prevent confusion with those of any other gull. (*See* Plate X.)

*Time.*—May and June.

*Remarks.*—Resident, but subject to much local movement. Notes: *kitt-aa*, *kitti-aa*, which sound like "get away, get away," or "Kittiwake, Kittiwake." Local and other names: Tarrock, Annet, Hacklet, or Hacket Gull, Waeg, Mackerel Bird. Gregarious, and a close sitter.

---

## LANDRAIL. *See* CRAKE, CORN.

---

## LAPWING. *Also* PEEWIT.

(*Vanellus vulgaris.*)

Order LIMICOLÆ; Family CHARADRIIDÆ (PLOVERS).

LAPWING ON HER NEST.

*Description of Parent Birds.*—Length about twelve inches. Beak somewhat short, straight, and black. Irides hazel. Forehead, crown, and long, narrow, upturned crest black, glossed with green. From the base of the beak a dirty white line passes over the eye and round under the crest. A streak of black runs under the eye and round to the nape of the neck, which is brown, mixed with white. Back and wing-coverts and scapulars brownish-green, glossed with purple and blue. Primaries black, with white spots at the ends of the first three or four; secondaries white on the basal half. Upper tail-coverts reddish-chestnut. Basal half of tail white; lower half black. Chin, throat, and upper half of breast black; lower half of breast, belly, and vent white; under tail-coverts pale rust colour. Legs and toes dull fleshy pink; claws black.

The female has the crest much shorter, is less bright in her colour, has whitish mark in black part of breast, and has a hoarser note.

*Situation and Locality.*—On the ground in rough pasture-land, marshes, fallow fields, and other suitable places throughout the British Isles. Although considerable flocks may be seen up and down the country in the winter time, there can be little doubt but that the species is diminishing in numbers owing to the collection of its eggs for table use. During the last few years I have seen old birds flocked in June, proving that from one cause or another they have been unable to rear young ones for that season.

*Materials.*—A few bits of dry grass, rushes, or moss, used as a lining to the depression in which the eggs are laid.

*Eggs.*—Four, although five have been reported. The latter number must be very rare, for I have found a great many nests, but never once saw more than four; and I know several gamekeepers who collect eggs for table use every spring, and they do not recollect ever meeting with a nest containing more, although my friend Mr. Bentham, of Oxted, found a nest near Redhill a year or two ago with five eggs in it. I have on several occasions found birds sitting hard upon only three eggs, and it is said that in such cases they frequently add a small stone or hard clod of earth to fill up the nest. Dirty olive-green, blotched and spotted all over with blackish-brown. Sometimes the ground-colour is light buff or buffish-brown, of various shades. The markings are generally most

LAPWING'S NEST AND EGGS.

numerous round the larger end. Size about 1.85 by 1.35 in. (*See* Plate VII.)

*Time.*—April, May, and June.

*Remarks.*—Resident, though subject to southern movement in winter. Notes: *peewit,* the first syllable long-drawn when used as a call-note, and short when the bird is alarmed. Spring call-note: *will-o-wit pee-weet.* Local and other names: Peewit, Green Plover, Peeweep, Tufit, Crested Lapwing. Sits lightly. A great deal has been written by some observers to prove that the bird rises straight off its eggs, and by others that it runs for some distance before taking wing. My own experience has been that the bird is exceedingly quick of eye and ear, and if the intruder be discovered at some distance the bird runs before rising; but if suddenly alarmed at close quarters it will rise straight off its nest. A good way to find the Lapwing's eggs is to creep very quietly up behind the wall or hedge of a field or pasture in which the birds are known to breed, and then show oneself suddenly, and mark where the birds rise from.

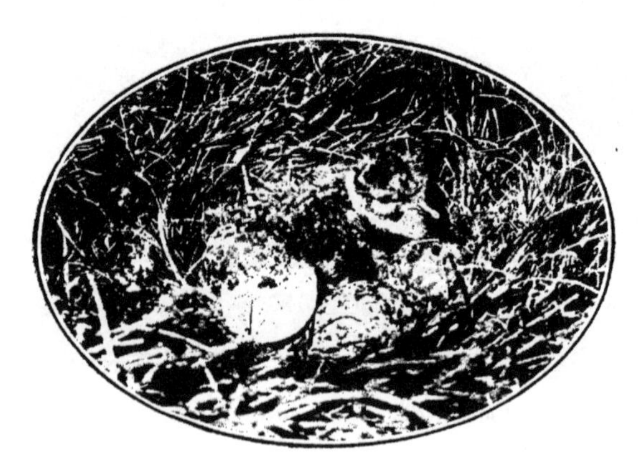

NEWLY HATCHED LAPWING AND EGGS

## LARK, SKY. *See* SKYLARK.

---

## LARK, WOOD. *See* WOODLARK.

---

## LINNET.

(*Linota cannabina.*)

Order PASSERES; Family FRINGILLIDÆ (FINCHES).

LINNET.

*Description of Parent Birds.*—Length about five and a half inches. Bill short, broad at the base, sharp-pointed, and bluish-grey in colour. Irides hazel. The plumage is subject to great variation. Forepart and top of head brownish-red; rest of head, back, and sides of neck brownish-grey; back and upper wing-coverts deep rufous-brown; wing-quills dusky, edged with white; upper tail-coverts dark brown; tail-quills brownish-black, edged with white, except the two centre feathers. Chin and throat and under-parts dirty reddish-brown, the red being brightest on the breast and very variable in its intensity, some having very little of it present. Legs, toes, and claws brown.

The female is a trifle smaller, lacks the red on the top of the head and breast, the feathers on her head, neck, back, and rump being dark brown, edged with a paler tint of the same colour; her under-parts are dull yellowish-brown, streaked with dark brown. She is said to have less white on the wing-quills.

LINNETS' NESTS AND EGGS.

*Situation and Locality.*—White and black thorn bushes, furze bushes, heath, juniper bushes, amongst tall heather. It is sometimes situated ten or twelve feet from the ground, and at others even on the ground, as represented in our first illustration. The nest is met with on furze-clad sides of hills, commons, and rough uncultivated lands covered with heather, furze, and ling, throughout the British Isles.

*Materials.*—A few small twigs, fibrous roots, dry grass-stems, moss, and wool, with an inner lining of hair and feathers, sometimes with rabbit or vegetable down.

*Eggs.*—Four to six, greyish-white, slightly tinged with blue or green, speckled, and spotted with purple, red, and reddish-brown; the spots are generally most numerous round the larger end of the egg. They closely resemble those of several other birds, such as the Greenfinch, Goldfinch, and Twite, and can only be distinguished with certainty by watching the parent birds on to or off their nests. Average measurement about .72 by .52 in. (*See* Plate II.)

*Time.*—April, May, and June, sometimes as late as July, and even August.

*Remarks.*—Migratory in bulk, but resident in small numbers. Notes: song, soft and low, mixed with some sweet and shrill notes. Local and other names: Brown Linnet, Common Linnet, Grey Linnet (a name referring to the young bird before the first moult), Whin Linnet, Red-breasted Linnet, Rose Linnet, Greater Redpole, Red Linnet, Linnet Finch, Red-headed Finch, Lintie, Linwhite. Sits closely.

## LINNET, MOUNTAIN. *See* TWITE.

---

## MALLARD. *See* DUCK, WILD.

---

## MAGPIE.

(*Pica rustica.*)

Order PASSERES ; Family CORVIDÆ (CROWS).

MAGPIE'S NEST AND EGGS.

*Description of Parent Birds.* — Length about eighteen inches, more than half of which is accounted for by the bird's abnormally long tail. Bill of medium length, slightly curved downward, and black. Irides hazel. Head, neck, back, wings, tail-coverts and quills black, glossed with green, purple, and blue, according to the light upon them. The scapulars, and part of the inner webs of some of the primaries, are white. The tail is very much wedge-shaped. Chin, throat, and upper breast black ; lower breast, belly, and sides white ; thighs and under tail-coverts black. Legs, toes, and claws black.

The female is not quite so large, nor is her plumage so brilliant as in the case of the male.

*Situation and Locality.*—In trees and thorn bushes at varying heights from the ground. Instances are on record of the bird having made its

MAGPIE'S NEST.

nest even in a gooseberry bush. It is sometimes partial to situations near the habitations of man, and a small clump of trees or thorn bushes will suit its purpose quite as well as a big wood or plantation, apparently, for I have as often found it in one as the other. Pretty generally over the British Isles, except Orkneys, Shetlands, and Outer Hebrides. A decreasing species, the diminution of its numbers being connected with its aptitude for devouring other birds' eggs and the consequent attention of the gamekeeper.

*Materials.*—Dead thorn-sticks, brambles, and twigs interlaced; those forming the foundation of the nest are firmly plastered together with liberal quantities of clay and mud. The nest is bulky, domed, and spherical, with a hole on the side and near the top. It is lined internally with fibrous roots, and is such a substantial affair that I once saw a gamekeeper shoot at one from the foot of a fir tree some fifty feet in height without breaking one of the six eggs inside.

*Eggs.*—Six to eight or nine, the first number being the most general. Dirty bluish-green, or yellowish-brown, spotted, freckled, and blotched all over with grey and greenish-brown. Size about 1.35 by .95 in. (*See* Plate I.)

*Time.*—March, April, and May.

*Remarks.*—Resident. Notes, chattering. Local and other names: Pyet, Madge, Mag, Maggie, Pianet, Hagister. I have known the bird sometimes sit closely when not far advanced in incubation, and at others lightly; something, I am inclined to think, depending upon the height of the nest from the ground.

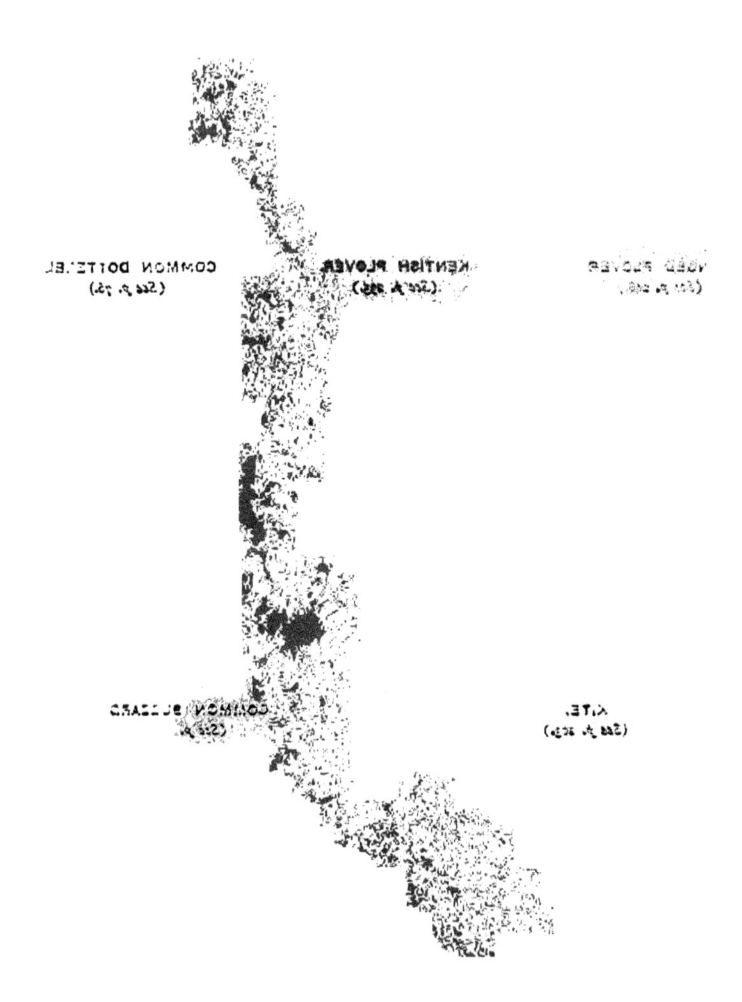

LAPWING.
(*See p.* 210.)

RINGED PLOVER.
(*See p.* 208.)

KE

KITE.
(*See p.* 203.)

NOTE.—*In referring to the eggs th*

PLATE 7.

## MARTIN. *Also* MARTIN, HOUSE.

(*Chelidon urbica.*)

Order PASSERES ; Family HIRUNDINIDÆ (SWALLOWS).

*Description of Parent Birds.*—Length about five and a quarter inches. Bill short, flat, and wide at the base, and black. Irides brown. Crown, nape, and sides of the head, back, and wing-coverts, dark steely-blue. Wing-quills dull black ; rump and upper tail-coverts white ; tail-quills dull black. The tail is forked, but so much less so than that of the Swallow that, apart from any difference of coloration, it easily distinguishes the bird. Chin, throat, breast, belly, and under-parts generally, white. Legs and toes short, and almost hidden by a profusion of fine, soft, white feathers. Claws grey.

The female is very similar indeed in appearance, although her plumage is perhaps not so bright.

*Situation and Locality.*—Under the eaves of houses, stables, barns, and other buildings, angles of windows, under the projecting "through" stones of barns ; in nooks and corners of rocks and sea cliffs. I know a small stable in Surrey, under the eaves of which, and principally on the side with a south-east aspect, I have counted forty-seven occupied nests several years in succession. My brother and I went down in 1894 specially to photograph the building, and to our great disappointment there was not a single nest under its eaves ! The owner informed us that it was the first time he had noticed the absence of the Martins for twenty-five years, and attributed it to the droughty summer of 1893 having made suitable building materials difficult to procure, and the unbearable persecution and robbery of the Sparrows. At the time of revising this work (1906) the Martins have never come back to the above-mentioned favourite haunt, and there is every reason for believing that the species is decreasing all

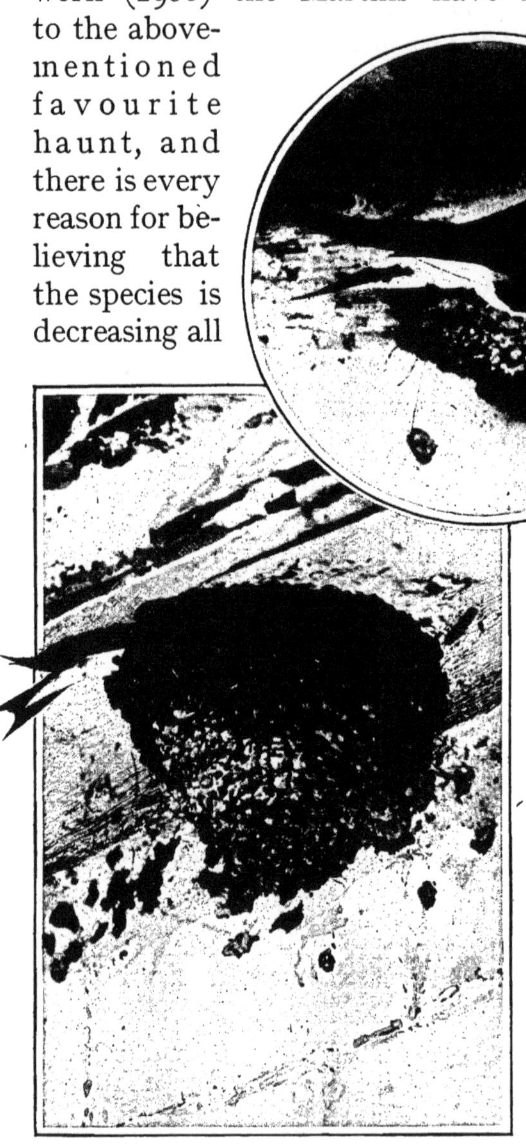

HOUSE MARTIN BUILDING.

HOUSE MARTIN ENTERING NEST.

over the country. General over the British Isles.

*Materials.*—Clay or mud made into pellets and cemented together until they form a kind of shell like the half of a deep basin, fixed close up under an eave or projecting object, with an elliptical hole at the top, and generally on one side. It is lined with bits of straw, hay, and feathers.

*Eggs.*—Four or five, rarely six; white and unspotted, the yolk giving them a slight pinky tinge before they are blown. One or two observers have recorded in the *Field* the finding of rust-red spotted specimens; but I have never met with a single egg showing any inclination in this direction out of many scores of nests examined. Size about .8 by .52 in.

*Time.*—May, June, July, August, and even as late as September.

*Remarks.*—Migratory, arriving in April and leaving in September and October; although individuals are frequently reported in November and even December. Notes: call, something like *spitz*, but very difficult to represent. Local and other names: Window Martin, Window Swallow, Eave Swallow, Martlet. Gregarious, as a rule. Sits closely.

---

## MARTIN, HOUSE. *See* MARTIN.

---

## MARTIN, SAND.

(*Cotile riparia.*)

Order PASSERES; Family HIRUNDINIDÆ (SWALLOWS).

SAND MARTIN.

*Description of Parent Birds.*—Length about four and three-quarters to five inches. Bill short, slightly turned down at the tip, and brownish black. Irides hazel. Head, back of neck, and all upper parts, including wings and tail, brownish-black or mouse-colour.

Throat and breast white, the latter having a band of lightish brown running across it. Belly and under-parts white. Legs and toes reddish-brown. The brown tinge of the upper-parts and the smaller size readily distinguish this bird from the House Martin.

The female differs very slightly from the male.

*Situation and Locality.*—At the extremity of a tunnel, dug by the bird's own exertions. It varies in length from eighteen inches to three or four feet, and generally slopes upwards from the entrance to the little round chamber in which the nest is situated. The gallery is about two inches in diameter, and generally crooked. In the banks of rivers, sand pits, railway cuttings, and lanes with high, sandy banks. Our illustration is from a photograph taken near Nutfield, Surrey, and is of double interest—firstly, it is quite away from any water, and secondly, every nest was taken possession of in 1894 by House Sparrows. These little birds breed all over the country, penetrating to the distant Orkney and Shetland Islands.

*Materials.*—Straw and grass stems, with an inner lining of feathers. The whole is very loosely put together, and I have met with specimens with no feathers at all, and but very few straws whereon the eggs were laid.

*Eggs.*—Four or five, seldom six; pure white when blown. The shell is so thin and semitransparent that the yolk shows through and gives the egg a pinky tinge. Size about .7 by .48 in.

*Time.*—May, June, and July.

*Remarks.*—Migratory, arriving in this country in March and April, and leaving in September and October. Call-notes loud and harsh, something

SAND MARTINS' NESTING HOLES.

like *share*, according to Nauman, but very difficult to represent in letters. Local and other names: Pit Martin, Land Swallow, Bank Swallow, River Swallow, Bank Martin. Gregarious. Sits closely.

---

## MERGANSER, RED-BREASTED.

(*Mergus serrator.*)

Order Anseres ; Family Anatidæ (DUCKS).

RED-BREASTED MERGANSER ON NEST.

*Description of Parent Birds.*—Length about twenty-two inches. Bill rather long, sharp, straight, and red, except the upper part, which is brownish. Irides red. Head and a little of the upper part of the neck glossy green, the feathers on the back of the head being lengthened. A line of black runs from the back of the head down behind the neck to the upper part of the back, which is also black; lower part of back, rump, and upper tail-coverts grey. Tail-quills brownish-grey. Wings a mixture of dark brown, white and black on the upper parts; primaries brownish-black; middle half of sides of neck white; breast rusty-red, spotted with black on the front; on the sides, in front of wing-points or shoulders, are a few white feathers edged broadly with black. Breast, belly, and under tail-coverts white. Legs, toes, and webs deep orange, tinged with brown; claws black. At the latter end of

RED-BREASTED MERGANSER'S NEST AND EGGS.

May the head and neck turn from glossy green to dull brown, and the rusty-red on the breast disappears.

The female is slightly smaller; her head, neck, and the whole of her upper-parts are of varying shades of brown, with two white bars on the wings. Front of neck white, mottled with light reddish-brown; under-parts white. Both sexes subject to variation of colour.

*Situation and Locality.*—On the ground, under bushes, banks, projecting ledges of rock; amongst

YOUNG RED-BREASTED MERGANSERS.

heather and brambles; occasionally in holes in trees, in rabbit-holes, holes and crevices of rocks; on small islands in lakes, on the shores of lakes, generally not far from the water. In the north of Scotland, Orkneys, Shetlands, Hebrides, and in Ireland. Our illustrations were procured in the Highlands, where I have upon occasion found as many as three nests within thirty yards of each other on a small island.

*Materials.*—Dead grass, roots, and rushes, in scanty quantities, lined with tufts of down from the bird's own body. These are light greyish-brown, with pale centres and tips. Sometimes no materials whatever except down are provided.

*Eggs.*—Six or seven to eleven or twelve, olive-grey to buffish-grey. Size about 2.6 by 1.7 in.

*Time.*—End of May, June, and the beginning of July.

*Remarks.*—A winter visitor; numbers, however, stay and breed with us. Local and other names: Red-breasted Goose, Sheld Duck, and Spear Wigeon (the latter two names only applied to the bird in Ireland). A very close sitter. I have trodden upon a bird on her nest in the heather before she would move.

---

## MERLIN.

(*Falco æsalon.*)

Order ACCIPITRES; Family FALCONIDÆ (FALCONS).

MERLIN ON NEST.

*Description of Parent Birds.*—Length about ten inches. Beak short, much curved, and bluish-horn colour. Bare skin round the base of the beak yellow. Irides dark brown. Crown bluish-grey, marked with black streaks along the shafts of the feathers. Cheeks and upper part of neck rusty-brown, marked with blackish streaks; back, scapulars, wing-coverts, and rump bluish-grey, each feather having the shaft black; wing-primaries black. Tail-quills, like the back, barred with a darker hue, and tipped with white. Chin and throat nearly white; breast, belly, sides, and thighs rusty-red, streaked with dusky-brown; vent and under tail-coverts pale rust-colour. Legs and toes yellow; claws black.

The female is about two inches longer than the male. The whole of her upper parts are dark liver-brown, the feathers tipped with rusty-red and having dusky shafts. Tail-quills, like the back, barred with light yellowish-brown. Under parts pale brownish-white, with broad, dusky-brown streaks.

*Situation and Locality.*—On the ground, amongst deep heather and ling or scattered rocks; on open moors, heaths, and rough sheep-pastures. It is said to be occasionally found in trees and is sometimes situated on rough heather slopes at the tops of cliffs. In the wild moorland parts of the north of England, Wales, Scotland, and Ireland. Our full-page illustration is from a photograph taken on the hills between Westmorland and Yorkshire. The nest was in deep heather on a sloping hillside, commanding every aspect of approach. It was evidently a favourite site, for the gentleman who showed it to us in 1894 said that a brood had been reared at the same place the year before; and I flushed the female close to the place in 1895, but was unable to find her nest. He showed us three knolls, each about fifty yards from the nesting site, upon which the old birds plucked the prey before taking it to their young. They brought Moor Poults (young Grouse), Green Plovers, Meadow Pipits, mice, and young Snipe.

*Materials.*—A few twigs or sprigs of heather, grass, or moss, generally next to nothing. The one photographed was in a very slight depression, and contained only a few dead heather sprouts.

*Eggs.*—Three to six, generally four or five,

MERLIN'S NEST AND EGGS.

creamy-white, so thickly covered with spots, blotches, or clouds of dark reddish-brown, as to almost completely hide the ground-colour. Sometimes the markings consist of small dots closely and thickly scattered over the whole surface, and in these the ground-colour becomes more apparent. Only distinguishable from those of the Hobby and Kestrel by the situation of the nest. Size about 1.6 by 1.2 in. (*See* Plate V.)

*Time.*—May and June.

*Remarks.*—Resident, but subject to a southern migration in October. Notes: a shrill, chattering cry. Local and other names: Blue Hawk, Stone Falcon. Sits lightly, in some cases, but closely in others. The bird figuring in the initial cut to this article was so bold that I could not frighten her off her eggs until I crawled from my hiding tent only a few feet away. This our smallest member of the Falcon family is an exceedingly plucky little bird, but unfortunately its numbers have to be kept down on grouse moors on account of the damage it does amongst the young birds.

YOUNG MERLINS IN THE NEST.

## MOORHEN. *Also* WATERHEN.

(*Gallinula chloropus.*)

Order FULICARIÆ ; Family RALLIDÆ (RAILS).

MOORHEN ON ICE.

*Description of Parent Birds.*—Length about thirteen inches. Bill of medium length, nearly straight, greenish-yellow at the tip, and red at the base and for some distance up the naked forehead, where the coot is white. Irides reddish-hazel. Head and neck dark bluish-grey. Back, wings, rump, and upper side of tail dark olive-brown. Breast and sides dark bluish-grey; belly and vent grey; flanks streaked with white. Under tail-coverts white. Legs and feet greenish-yellow; claws rather long and brown. Above the knee

MOORHEN'S NEST AND EGGS.

is a broad garter of red. The female is rather larger, and brighter in the coloration of her plumage, than the male.

*Situation and Locality.*—Generally on the ground, amongst flags, rushes, reeds, low bushes growing from the water; reeds and coarse aquatic plants growing in and on the banks of rivers, small streams, canals, ponds, lakes, and reservoirs. I have, however, met with it at considerable heights above the water, amongst rubbish left by an abnormally high flood in a tree. Common throughout England, Wales, Scotland, and Ireland.

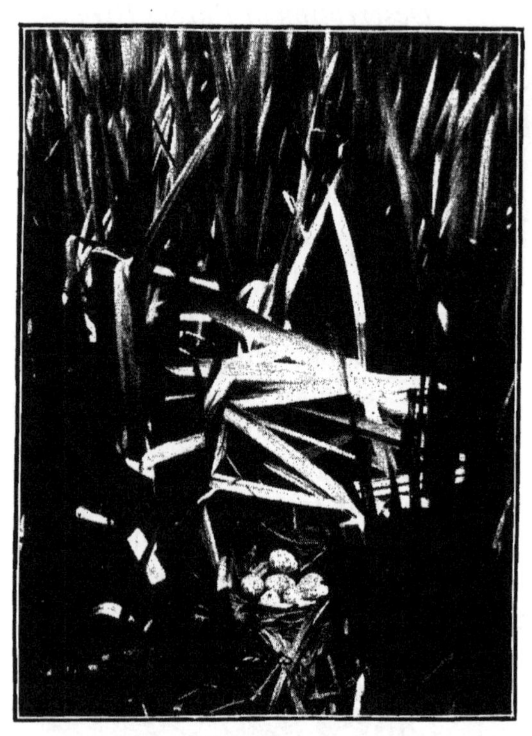

MOORHEN'S NEST AND EGGS.

*Materials.*—Flags, reeds, rushes, and grass. Generally in small quantities where the situation is dry, but often in a fair-sized matted mass where its base is in the water.

*Eggs.*—Seven to ten. Buffish-white or rusty-buff, spotted and speckled with reddish-brown of various shades. The markings are not very large, or profusely distributed. Size about 1.7 by 1.2 in. (*See* Plate XIV.)

*Time.*—March, April, May, June, July, and sometimes even as late as August.

*Remarks.*—Resident, and partially migratory. Notes: Call, *crek-rek-rek.* Local and other names: Water Hen, Marsh Hen, Moat Hen, Gallinule. Not a very close sitter, slipping quietly off the nest and instantly hiding on the approach of any intruder. The eggs may be distinguished from those of the Coot by their smaller size and larger markings.

---

## NIGHTINGALE.

(*Daulias luscinia.*)

Order PASSERES ; Family TURDIDÆ (THRUSHES).

NIGHTINGALE.

*Description of Parent Birds.* — Length about six inches. Bill of medium length, nearly straight, and brown. Irides hazel. Head and upper parts of body uniform tawny-brown. Wing- and tail-quills brown, edged with rust-colour. Chin, throat, and all under-parts greyish-white, tinged with brown on the breast, and reddish on the under tail-coverts. Legs, toes, and claws brown.

The female is rather smaller in size, but otherwise closely resembles the male.

*Situation and Locality.*—In natural declivities on

the ground, on little banks at the foot of trees, amongst exposed roots at the bottom of hedgerows, under the shelter of ferns or weeds. I have known one at some little height from the ground, amongst the dead weeds, twigs, and leaves that had been built up round the trunk of a tree for the purpose of hiding a gunner in pursuit of Wood Pigeons. In woods, groves, small shady copses, plantations, quiet gardens, and commons with clumps of hazel briars and brambles growing thereon. It is peculiarly limited in its habitat, as a rule going no farther north than Ripon in Yorkshire, and no farther west than the Valley of the Exe, although individuals have been met with beyond these limits, and there is reason to believe it is extending its range.

*Materials.*—Dry grass-stalks, leaves, moss, bits of bark and fibrous roots, lined inside with fine grass and horsehair.

*Eggs.*—Four to six, generally five. Uniform olive-brown or olive-green. Specimens have been met with occasionally of a greenish-blue colour, minus any reddish-brown, which colour upon the greenish-blue produces olive-brown. Sometimes the brown is disposed in a kind of cap at one or other of the ends of the egg, or in streaks. Size about .85 by .6 in. (*See* Plate IV.)

*Time.*—May and June.

*Remarks.*—Migratory, arriving in April and leaving in August. Notes: Call, *purr, purr,* and something like *wheet!* Bechstein rendering it *witt, krr,* and adding that it has a note of enjoyment represented by a deep *tack.* Its song is uttered both by day and night, and is ravishingly sweet and melodious. Local or other name: Philomel. Sits closely, and slips away without demonstration.

NIGHTINGALE'S NEST AND EGGS.

## NIGHTJAR. *Also* GOATSUCKER.

(*Caprimulgus europæus.*)

Order PICARIÆ; Family CAPRIMULGIDÆ (GOATSUCKERS).

YOUNG NIGHTJARS.

*Description of Parent Birds.*—Length about ten and a half inches. Bill very short, upper mandible slightly turned downward, flexible, and dusky black. The gape is very wide, and furnished on the upper side with a number of stiff bristles. The plumage on the upper part of the body consists of a beautifully diversified mixture of brown, black, rusty-red, and white, spotted, and sprinkled with grey. The under-parts greyish and rusty-brown, barred and freckled with dark brown. There are a few white markings round the throat. On the three first quill-feathers of the wings is an oval white spot, also on the two outside feathers of the tail. Legs short, rough, scaly, and feathered to below the knee. Middle toe considerably longer than the rest, and the claw upon it is serrated on one side; all orange-brown. The female is somewhat darker, and lacks the white markings on wings and tail.

*Situation and Locality.*—On the ground beneath furze bushes, brackens, heather, or quite in the open on commons, heaths, open bramble-covered woods and copses in nearly all suitable districts throughout the British Isles, but not very numerous anywhere. I have met with it most frequently in the southern and eastern counties of England.

NIGHTJAR'S EGGS

*Materials.*—None, the eggs being laid on the bare ground.

*Eggs.*—Two; ground-colour white, greyish-white, or creamy-white, clouded, blotched, marbled, or veined with dark brown, and underlying tints of bluish lead-colour. They are subject to great variation, and often closely resemble flint pebbles one may pick up on the beach with chalk adhering to them; in fact, I have on more than one occasion been deceived by one of these pebbles lying under a furze bush. Size about 1.25 by .87 in. (*See* Plate V.)

*Time.*—May and June.

*Remarks.*—Migratory, arriving in May and departing in September or October. Notes: *jar-r-r-r-r-r* and *dee, dee,* said to be uttered on taking flight. Local and other names: Goatsucker, Dorhawk, Fen Owl, Nighthawk, Wheelbird, Jar Owl, Churn Owl, Goat Owl. Sits very closely, trusting to the wonderful harmony of her plumage with surrounding objects. When at roost in trees it generally perches along and not across a branch, thus considerably enhancing its chances of escaping detection.

NIGHTJAR COVERING YOUNG.

## NUTHATCH.

(*Sitta cœsia.*)

Order PASSERES ; Family SITTIDÆ (NUTHATCHES).

NUTHATCH.

*Description of Parent Birds.*—Length about six inches ; bill moderately long, strong, nearly straight, sharp at the tip, and bluish-black, except at the base of the lower mandible, where it is whitish. Irides hazel. Crown and all upper parts of body, including wing-coverts and part of tail, bluish slate-grey. Wing-quills dusky, margined on the outer webs with blue. Tail-quills, excepting those mentioned above, black, tipped with grey, and marked on either side with white. A black streak passes from the base of the bill to each eye, and thence down the side of the neck. Sides of head and chin white ; throat, breast, and belly buff ; sides and thighs dark rust-colour or chestnut ; vent white, marked with rust-colour. Legs, toes, and claws light brown, inclining to yellowish.

The female is lighter coloured on her under-parts.

*Situation and Locality.*—In a hole in the trunk or strong branch of a tree, old stumps, and occasionally in a hay rick or wall ; at varying depths of from three or four to twelve or fifteen inches. Breeds generally throughout England, although most numerously in the wooded parts of the south, eastern and midland counties of England, also

NESTING HOLE OF NUTHATCH.

Wales; rarely met with in Scotland and never in Ireland.

*Materials.*—Leaves, flakes of bark, dry grass, and sometimes chips and *débris* when the bird is obliged to enlarge the situation selected. The bird has the peculiar habit of plastering up the approach to its nest with clay if there be more room than is necessary for its admission.

*Eggs.*—Five to eight or nine; pure white, spotted with reddish-brown; sometimes blotched, the markings varying in distribution. If care is not exercised the eggs are likely to be mistaken for those of the Great Titmouse, but the character of the nest will readily settle the point. Size about .8 by .57 in. (*See* Plate III.)

*Time.*—April, May, June, and July.

*Remarks.*—Resident. Notes: call, *whit, whit, whit;* sometimes represented as *twi-twit, twi-twit.* Local and other names: Woodcracker, Nutjobber, Jarbird, Nuthack, Mudstopper. Sits very closely, and hisses like a snake when disturbed.

---

## ORIOLE, GOLDEN.

(*Oriolus galbula.*)

Order PASSERES; Family ORIOLIDÆ (ORIOLES).

This bird, although a somewhat rare and accidental visitor to our shores, has, according to some authorities, bred in several parts of England. There are, however, sceptics who doubt this, and adduce, as a reason, that there is not a collector who can boast the possession of a British-laid specimen. Be this as it may, it is doubtful whether the bird will ever succeed in breeding in this country, on account of the eagerness with which the collector seeks after the skin of the male, whose attractive colours excite his cupidity.

---

## OSPREY.

(*Pandion haliaëtus.*)

Order ACCIPITRES; Family FALCONIDÆ (FALCONS).

OSPREY.

*Description of Parent Birds.*—Length about twenty-two inches. Beak short, much curved, and black; naked skin round base of beak blue. Irides yellow. Crown and nape whitish, streaked with dark brown, the feathers somewhat elongated into a kind of crest. Back and wings dark brown, sometimes glossed with purple, ends of the latter black. Tail waved with two shades of brown; chin and throat white, shaded with light brown across the breast; belly, sides, thighs, and under tail-coverts white. Legs and toes blue, claws long, strong, much curved, and black.

The female is slightly larger, and has her head and breast more marked with brown, according to Mr. Seebohm; although this difference does not appear very noticeable, so far as my experience goes, when studying the birds at home amidst their natural surroundings.

*Situation and Locality.*—Near the top of a tree, the summit of an inland crag, or on the highest point of some ruin upon an island or commanding promontory amongst the lonely lochs of the Highlands of Scotland. It is only known to breed in

AN OSPREY AND ITS EYRIE

one or two counties (Inverness-shire, Ross-shire, and Galloway), and I regret to state that at the time of revising this work (1906) two of the eyries I have known for a long time are now quite deserted.

*Materials.*—Sticks, twigs, turf, moss, and grass. The structure is of a huge character, and the top almost flat. The same site is used again and again with the utmost regularity.

OSPREYS AT HOME ON A HIGHLAND LOCH.

*Eggs.*—Two to four, generally three; very variable and beautiful. The ground-colour ranges from white to dull yellowish-white, handsomely marked with rich reddish-brown and light brownish-grey. Some examples are suffused with bright orange-red or purple. The blotches and spots are sometimes very thickly distributed, at others they form a zone round the larger end or are irregularly scattered over the entire surface of the egg. The eggs also vary considerably in size. Average about 2.3 by 1.85 in. (*See* Plate VI.)

*Time.*—May and June.

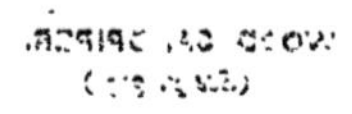

(*See p.* 65.)

WOOD SANDPIPER.

(*See p.* 339.)

R
P
(

OYSTER-CATCHER.

(*See p.* 261.)

NOTE.—*In referring to the eggs*

PLATE 8.

*Remarks.*—Migratory, arriving in April or May, and departing in September and October. Notes: *kai, kai, kai.* Local and other names: Eagle Fisher, Mullet Hawk, Fish Hawk. Sits lightly, according to Mr. Dixon, but pretty closely according to Mr. Seebohm. My own experience coincides with that of the former authority.

---

## OUZEL, RING.

(*Turdus torquatus.*)

Order PASSERES; Family TURDIDÆ (THRUSHES).

YOUNG RING OUZEL.

*Description of Parent Birds.*—Length about eleven inches. Bill brownish-black, with a variable amount of yellow at the base, nearly straight, and of medium length. Irides hazel. Head, neck, back, wings, rump, and tail black, slightly tinged with brown, and margined, more or less, with grey, especially on the wings. Chin, throat, and underparts blackish-brown, the feathers being bordered with grey. Across the breast is a broad, curved band or crescent of white, edged with a brownish tint. The legs, toes, and claws are brownish-black.

The female is browner and greyer, and the crescent on her breast much duller and less defined.

*Situation and Locality.*—In clefts and on ledges of rock, steep banks, holes in stone walls, barns, limekilns, and sometimes quite on the ground. In the summer of 1907 I found a nest in an isolated holly bush in Westmorland. It was some ten to twelve feet from the ground. In the mountain

RING OUZEL'S NEST AND EGGS.

and moorland parts of the north and west of England, Scotland, Wales, and Ireland. The illustration on page 244, procured in Westmorland, is a very typical example of the situation of the nest.

*Materials.*—Small twigs, roots, coarse grass, moss, and mud, with an inner lining of fine grass. It is a very similar structure to that of the Blackbird.

*Eggs.*—Four or five, blue-green, freckled, and spotted with brown. They are, as a rule, covered with larger spots than the eggs of the Blackbird, but upon occasion the latter will lay eggs resembling them so closely that it is quite impossible to distinguish without seeing the parent birds or knowing something of the locality of the nest. Size about 1.2 by .84 in. (*See* Plate IV.)

*Time.*—April, May, and June, generally the last two months.

*Remarks.*—Migratory, arriving in April and departing about the end of October. Notes: song, desultory, plaintive, and far-sounding. Local and other names: Rock Thrush, Ring Thrush, Rock Ouzel, Tor Ouzel, Ring Blackbird. Sits pretty closely, and is demonstrative when disturbed.

MALE RING OUZEL FEEDING FEMALE ON NEST.

## OUZEL, WATER. *See* DIPPER.

---

## OWL, BARN. *Also* SCREECH OWL.

(*Strix flammea.*)

Order STRIGES; Family STRIGIDÆ (OWLS).

BARN OWL.

*Description of Parent Birds.*—Length about fourteen inches. Beak short, much curved at the point, and pale grey. Irides black. Discs round the eyes white, with the exception of a little patch close to the eyeball on the inner side of each, which is rufous. The feathers, especially on the lower outside of each disc, are tipped with light rusty-brown of varying shades. Crown and nape light buff, sprinkled with grey and spotted with dark brown and dirty white. Back, wings, and rump buff, with a lacework of grey, on which are more or less perpendicular lines of spots of dull black and dirty white; upper surface of tail-quills greyish-buff, crossed by fine darkish-grey bars; upper breast white, slightly tinged with buff; lower breast, belly, vent, and under tail-coverts white. Legs covered with white downy feathers, toes with short hairs; claws brown.

The female is distinguished by a few dark brown spots on her sides and belly.

*Situation and Locality.*—In hollow trees, church towers, barns, pigeon-cotes, crevices of rocks overshadowed by ivy, and old ruins; pretty generally

HOLLOW TREE IN WHICH A PAIR OF BARN OWLS HAVE NESTED FOR SEVERAL YEARS.

throughout the British Isles, but scarce in the Highlands, Orkneys, and Shetlands. I have found more eggs and young belonging to this species in hollow trees than any other kind of situation. Our full-page illustration represents a hollow tree at Shenley, in Hertfordshire, in which a pair of Barn Owls have nested for many years.

*Materials.*—None usually, except the pellets of undigested parts of small birds and mice; how-

YOUNG BARN OWLS

ever, in some situations a few sticks or twigs and other materials, such as straws, wool, and hair in small quantities, are said to be used.

*Eggs.*—Two to six; white, without polish or markings of any kind. The bird commences to sit as soon as she has laid one or two eggs, and keeps on laying at intervals as incubation advances, so that the young in the same nest may often be found in various stages of development. Average size about 1.6 by 1.25 in. Their smaller size distinguishes them from those of the Tawny Owl, and their situation from those of both species of Horned Owls.

*Time.*—April, May, June, and July, although young have been found as late even as December.

*Remarks.*—Resident. Note: a loud screech. Local and other names: White Owl, Hissing Owl, Church Owl, Madge Howlet, Jinny Oolet, or Oolert, Screech Owl, Yellow Owl. Sits closely, and is partial to old situations.

---

## OWL, LONG-EARED.

(*Asio otus.*)

Order STRIGES; Family STRIGIDÆ (OWLS).

LONG-EARED OWL.

*Description of Parent Birds.* — Length about fourteen inches. Beak short, much curved, and dusky horn colour. Irides orange-yellow. Facial discs dusky brown near the centre on the inside, and white towards the ends of the feathers; the outer sides of each disc are pale brown, ending in a line of darker brown. The ears or tufts of feathers on the head are about an inch and a half long, greyish-white on the inner edges, brownish-black in the centre, and dullish yellow on the outside edges. Crown, between horns, a mixture of the same colours. Nape, neck all round, and the upper portion of the back, dull yellow, streaked, longitudinally, with brownish-black. Lower back and wings yellowish-brown, marked with greyish-white, dark brown, and black. Upper side of tail rusty-red, barred and speckled with dark brown; breast and belly greyish-white, mixed with pale brown, and streaked and barred with dark brown. Under tail-coverts, and feathers on legs and toes, pale yellowish-brown; claws same colour as beak.

The female is similar in plumage, but is somewhat larger.

*Situation and Locality.*—The old nest of a Crow, Heron, Magpie, Wood Pigeon, or the disused drey of a Squirrel, in plantations of firs, and in woods and forests containing evergreens, sparingly in suitable localities throughout the United Kingdom. In woods where none of the creatures above mentioned breed the Long-Eared Owl lays her eggs upon the ground under heather, stunted firs, or some kind of cover. It has been stated that this species breeds in North Uist, but I do not think this is the case. I have met with it breeding in the Highlands, East Anglia, and the south of England.

*Materials.*—None.

*Eggs.*—Three to seven, generally four or five. White, oval, and smooth. Size about 1.65 by 1.3 in. Not likely to be confused with those of any other bird except Ring Dove; but their number and the appearance of the layer will readily settle the point.

*Time.*—March, April, and May.

*Remarks.*—Resident, and also migratory. Note, a deep hoot. When young ones are approached, the old birds, ever on the watch, sometimes begin to perform the strangest antics. Whilst in the Highlands on one occasion I was shown the young owls figured opposite by a gamekeeper. The parent birds resented my attentions to their offspring, and

LONG-EARED OWL'S NEST.

flying to and fro over my head smacked their wings together over and under their bodies, and alighting under a group of stunted firs, squealed like young rabbits in the throes of death, or mewed like cats in order to attract my attention. Local and other names, none. Sits very closely.

YOUNG LONG-EARED OWLS.

---

## OWL, SHORT-EARED.

(*Asio accibitrinus.*)

Order STRIGES ; Family STRIGIDÆ (OWLS).

*Description of Parent Birds.*—Length about fifteen inches. Beak short, much curved, and blackish. Irides yellow. The radiating circle of feathers round each eye black in the centre, and lighter on the outer edges, mixed with reddish-brown, black, and white, especially the last-named colour, round the bill. On the top of the head are two tufts of feathers about three-quarters of an inch long, which the bird can erect or depress at pleasure. These are blackish on the outer

SHORT-EARED OWL'S NEST AND EGGS.

webs, and whitish on the inner. Crown of the head, neck, back, and scapulars dusky, the feathers being bordered with light rusty-brown. Wing-coverts dusky, marked with a number of yellowish-white spots; primaries light rusty-brown, barred with blackish-brown. Tail-quills pale rufous, barred with dark brown. Under-parts buffish-white, streaked on the breast and belly with blackish-brown. Legs and toes feathered and pale
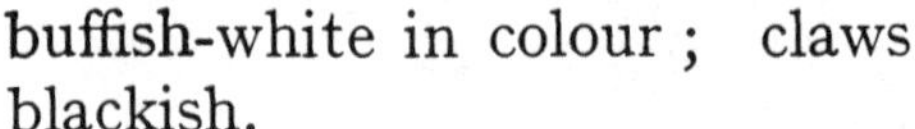
buffish-white in colour; claws blackish.

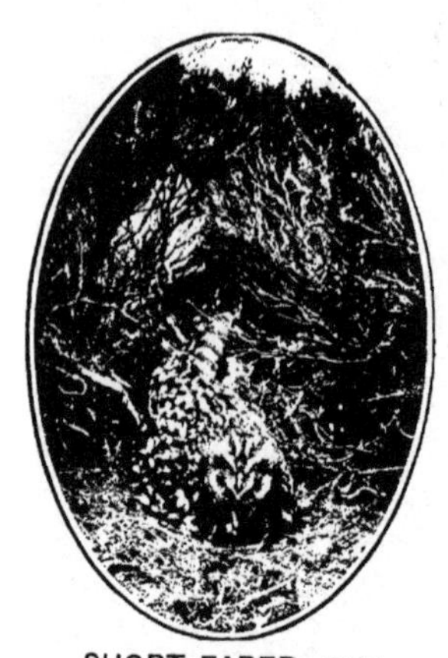
SHORT-EARED OWL ON NEST.

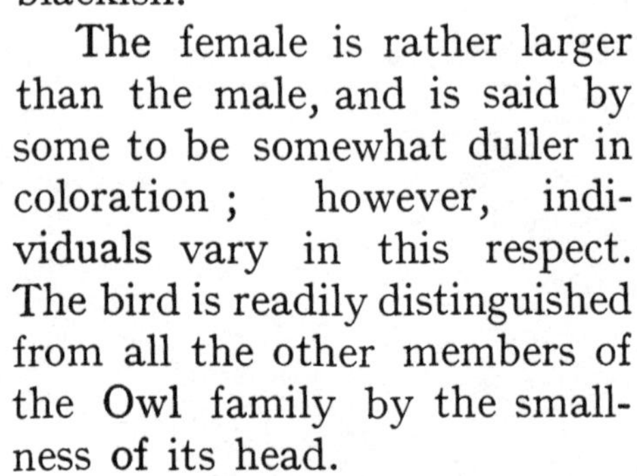
The female is rather larger than the male, and is said by some to be somewhat duller in coloration; however, individuals vary in this respect. The bird is readily distinguished from all the other members of the Owl family by the smallness of its head.

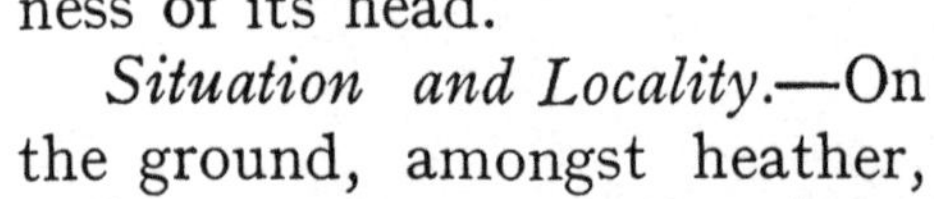
*Situation and Locality.*—On the ground, amongst heather, long grass, rushes, sedge, and gorse; on large moors, upland heaths, fens, and marshes in Norfolk, Suffolk, Cambridgeshire, and in the northern counties of England, Wales and Scotland. It is said to be only a winter visitor to Ireland.

*Materials.*—Dry grass, moss, and other bits of dead vegetation, used sparingly to line the hollow made or selected; sometimes none whatever.

*Eggs.*—Three to five, generally; sometimes as many as seven or eight. I have seen a clutch of seven in Norfolk, but never more than five in the Highlands. Mr. Richard Bell says that during the great Vole plague of 1890–93 one of his shepherds found a Short-eared Owl sitting upon twelve eggs on February 29th, 1892. The ground was covered with snow at the time, and seventeen dead Voles were lying upon it near the nest. This is very early for the species to breed, and, curiously enough, Mr. W. H. Hudson, in his "Naturalist in La Plata," says that during a plague of mice in South America Short-eared Owls bred in the middle of winter. White, and oval in form. Size about 1.6 by 1.28 in. Easily distinguished by nesting site.

*Time.*—April and May.

*Remarks.*—Resident and migratory, its numbers being swollen in winter by the arrival of more northern breeders. Notes: a shrill cry and snapping of the beak when the nest or young are in danger. Local and other names: Woodcock Owl, Hawk Owl, Mousehawk, Short-horned Howlet, Horned Oolert. A close sitter. During dull days I have frequently seen the old birds hunting for food for their young ones, both on the Norfolk Broads and in the Outer Hebrides.

YOUNG SHORT-EARED OWL AND EGGS.

## OWL, TAWNY. *Also* Wood Owl.

(*Syrnium aluco.*)

Order Striges ; Family Strigidæ (OWLS).

TAWNY OWL.

*Description of Parent Birds.*—Length about fifteen inches. Beak short, much curved, and horn white. Irides dark brown. The circle surrounding each eye is greyish-white, margined by a line of dark brown. Head, neck, back, and wings tawny-brown, finely marked with dark brown and black, and mixed with ash-grey. On the wing-coverts and scapulars are two descending lines of large white spots; the primaries are also barred with dark brown and dull white. Tail, two centre feathers uniform tawny-brown, rest barred with tawny- and dusky-brown. Breast, belly, and under parts greyish-white, streaked and mottled with two shades of brown. Under-coverts of tail white. Legs and toes covered with greyish-white feathers; claws large, much hooked, and horn white, with black tips.

The adult female is similar in plumage, but somewhat larger in size.

*Situation and Locality.*—The favourite nesting site is in a hole of a hollow tree, although the bird sometimes uses clefts of rocks, holes in the walls of stables and barns, deserted nests of Rooks, Magpies, Crows, and Hawks; also rabbit-burrows. This Owl is a lover of woods, forests, and parks, and is pretty generally scattered over England,

TAWNY OWL'S EGGS.

R

Wales, and the south of Scotland; rarer in the north, and absent from Ireland. I have met with it most numerously in Cumberland.

*Materials.*—None, the eggs being laid on decayed wood, old hay, or the bird's own "castings."

*Eggs.*—Three or four. Pure white, smooth, and round. Size about 1.8 by 1.52 in. *D*istinguishing features, the round shape and large size.

YOUNG TAWNY OWL.

*Time.*—March, April, May, June, July, August, September, October, and even later.

*Remarks.*—Resident. Notes: *tu-whit, to-whoo;* when pleased the bird utters a low kind of whistle, and when angered snaps its beak with considerable sound. Local and other names: Brown Owl, Wood Owl, Jinny Oolert, Hoot Owl, Jenny Howlet, Ivy Owl. Comes forth at night and hoots weirdly. I have on several occasions, however, heard it uttering its note by day. This bird is said to stand the light of day worse than any other member of the Owl family, although I have seen it abroad on dull days at noon, and even quartering a hedgerow in weak October sunshine. I have heard it hooting in the middle of a summer's day, when the sun was blazing high overhead, both in Westmorland and in the Highlands of Scotland. A close sitter.

---

OWL, WOOD. *See* OWL, TAWNY.

## OYSTER-CATCHER.

(*Hæmatopus ostralegus.*)

Order LIMICOLÆ ; Family CHARADRIIDÆ (PLOVERS).

YOUNG OYSTER-CATCHER.

*Description of Parent Birds.*—Length about sixteen inches. Bill long, straight, and orange-coloured. It is shaped like a vertical wedge, a form which renders it eminently useful for dislodging limpets and other univalves from rocks. Irides crimson. Head, neck, back, and wings black, with the exception of a white, broad, slanting bar across the last. Rump and upper half of tail white, lower half black. There is a small patch of white under the eye. Throat and upper part of breast black. Lower breast and all the under-parts of the body white. Legs and toes purple or bluish flesh colour ; claws black. In the early spring the bird wears a white collar or gorget on the neck, but this disappears as the season advances.

*Situation and Locality.*—On the ground, amongst the shingle and sand of the sea-shore. Pretty generally in suitable localities round our coasts, and frequently found quite inland on the banks of rivers and lakes. It is most numerous in Scotland and the surrounding isles. In the Outer Hebrides I have known a pair of birds use the same nest three years in succession.

*Materials.*—A few shells or pebbles are often used as a kind of pavement. Sometimes a few

OYSTER-CATCHER'S NEST AND EGGS.

bents are employed, or the eggs are laid on drifted seaweed.

*Eggs.*—Two to four, usually three. I have never seen a clutch of four myself. Yellowish, stone, or cream colour, streaked, blotched, and spotted with dark brown and grey. Occasionally the markings are inclined to form a zone at the larger end, but generally they are pretty evenly distributed over the shell. Size about 2.2 by 1.5 in. (*See* Plate VIII.)

*Time.*—May and June.

*Remarks.*—Resident. Notes: a clamorous chattering when the nest is approached. Local and other names: Sea Pie, Olive, Mussel Picker, Pienet, Tirma, Sea Piet, Trillichan, Chaldrick, Scolder, Sheldraker or Skelderdrake. Sits lightly, and generally has intimation of the approach of an intruder given by the male, which also assists in work of incubation, as I have proved by watching the birds take turn and turn about on the eggs from a stone hiding house built within ten feet of a nest in the Outer Hebrides.

OYSTER-CATCHER AND NEST.

## PARTRIDGE, COMMON.

(*Perdix cinerea.*)

Order GALLINÆ; Family PHASIANIDÆ (PHEASANTS).

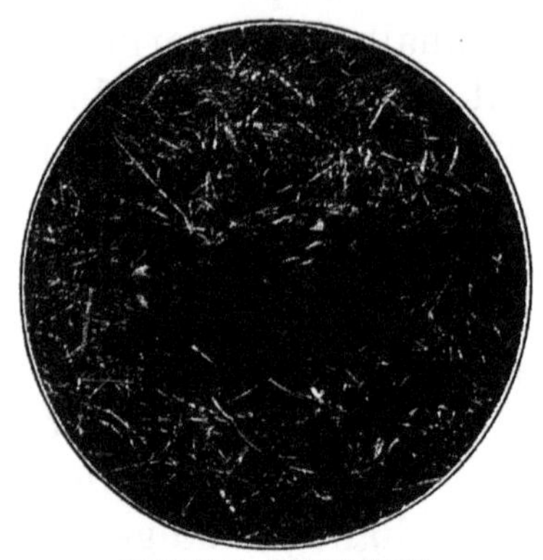
COMMON PARTRIDGE.

*Description of Parent Birds.* — Length about twelve and a half or thirteen inches. Beak short, curved downwards, and bluish-grey. Irides hazel. Forehead and cheeks bright rust-colour; behind the eye is a patch of naked red skin; crown, back of the neck, and shoulders, cinereous-brown; lower back and wing-coverts mottled with two shades of reddish-brown on a pale buff ground, the central line of each feather being pale buff, unmarked. Wing-quills brown, barred with pale buff. Tail-coverts and two centre quills brown, barred with rusty-red, remaining tail-feathers rusty-red; chin and throat bright rust-colour; neck, breast, and sides bluish-grey, freckled with a darker tinge of the same colour, marked on the breast with a horseshoe of rich bay, and barred on the sides with the same colour; under tail-coverts pale rusty-brown. Legs and toes bluish-grey; claws black.

The female is a trifle smaller than the male, and the rust-colour on her head is neither so extensive nor so bright, and the lesser and medium wing-coverts and scapulars are marked with buff cross-bars, a feminine distinction first observed by Mr. Ogilvie Grant.

COMMON PARTRIDGE'S NEST AND EGGS.

*Situation and Locality.*—On the ground at the bottom of a hedge, amongst mowing grass, clover, standing corn, weeds, brackens, rough grass, and brambles. In ploughed fields, pasture lands, on the outskirts of woods, and in grass-fields. Plentiful in cultivated districts, where preserved, but less numerous in high moorland districts, from which I have known the bird banished for years together by an exceptionally hard winter. In all suitable districts throughout the British Isles. Sometimes curious sites are chosen by this bird for its nest, such as on the thatch of a shed; and Booth mentions finding a Linnet's nest in the side of a stack, and that of a Partridge on the thatch of another close to it. I have known instances of the Common Partridge and Red-legged species laying in the same nest, also of a Red Grouse and Common Partridge joining in one.

*Materials.*—A few blades of dry grass, bits of bracken or dead leaves, used as a lining to the slight hollow selected.

*Eggs.*—Ten to sixteen or twenty; as many even as thirty-three have been recorded, but such a large number is undoubtedly the production of two hens. Pale olive-brown or greenish-yellow, unspotted. Size about 1.4 by 1.1 in. Easily distinguished from those of the Red-legged Partridge by smaller size, colour, and lack of spots. (*See* Plate XV.)

*Time.*—May and June, although nests containing fourteen eggs have been found as early as April 18th; and I have seen sitting hens have their heads cut off in grass-fields by the mower's scythe as late as the middle of July.

*Remarks.* — Resident. Notes: *turwit* (call); *ajick, jick* (alarm). Local or other name: none.

Sits very closely, and is of very uneven temper. Some individuals will suffer a great amount of intrusion, and others will forsake their eggs upon the slightest molestation. I have known a bird with twenty-four eggs trodden upon whilst sitting, and although five of her treasures were broken she came back and sat on the remainder.

COMMON PARTRIDGE ON NEST.

---

## PARTRIDGE, RED-LEGGED.

(*Caccabis rufa.*)

Order Gallinæ ; Family Phasianidæ (PHEASANTS).

RED-LEGGED PARTRIDGE ON NEST UNDER PLANT POT.

*Description of Parent Birds.*—Length about thirteen inches and a half. Bill short, curved downwards, and red. Irides red. Crown bright chestnut ; back of neck, back, rump, wing and tail-coverts, brownish ; wing-quills darker and tipped with light yellowish-brown; tail-quills chestnut and greyish-brown. A black streak runs from the nostrils to the eyes, then turns downwards, making a collar of black from which spots and streaks of the same colour extend towards the upper part of the breast. Breast pearly grey; belly, vent, and under tail-coverts fawn colour; sides and flanks transversely variegated with crescent-shaped marks of black, white, pearly-grey, and fawn colour. Legs and toes red, claws brown.

The female is not so large or bright and distinctive in coloration, and lacks the rounded knob which takes the place of a spur on the leg of the male.

*Situation and Locality.*—On the ground at the bottom of hedgerows, amongst tall grass and other herbage in corn, clover, and grass fields; occasionally it is said to select the thatch of a hayrick. In cultivated and uncultivated districts, such as commons and waste lands and heaths, more or less in all parts of England, but most plentifully in the southern and midland counties.

*Materials.*—Dry grass and dead leaves, used as a lining to the hollow selected.

*Eggs.*—Ten to eighteen; yellowish-brown or creamy-grey in ground-colour, spotted and speckled with reddish or cinnamon-brown. The spots vary in size and number, and the shell is coarse, pitted, and very strong. Average size about 1.55 by 1.2 in. (*See* Plate XV.)

*Time.*—April, May, and June.

*Remarks.*—Resident. It was introduced into this country about two hundred years ago, but has never gained a footing either in Scotland or Ireland. Notes said to resemble *cockileke.* They sound to me more like *dick-to-cher*, and by reproducing these

RED-LEGGED PARTRIDGE'S NEST WITH TWO PHEASANT'S EGGS IN IT.

sounds with a clay tobacco pipe I can call the birds to me. Local and other names: Frenchman, French Partridge, Guernsey Partridge. A close sitter.

---

PEEWIT. *See* LAPWING.

---

## PETREL, FULMAR.

(*Fulmarus glacialis.*)

Order TUBINARES; Family PROCELLARIIDÆ (PETRELS).

FULMAR PETREL.

*Description of Parent Birds.*—Length about nineteen inches. Beak of medium length, large, strong, nearly straight, with exception of the upper mandible, which is much curved downwards near the tip. It is of a yellowish colour, with a greenish tinge 'round the nostrils. Irides dark brown. Head and the whole of the neck white; back and wings French grey, except quills, which are darker; upper tail-coverts and tail-quills French grey; breast, belly, and under-parts white. Legs, toes, and webs pale grey. Many specimens are of an ash-grey or ash-brown tint all over, somewhat darker on the back and wings.

The female is similar to the male.

*Situation and Locality.*—On turf-covered shelves

and ledges; also in crevices of high, inaccessible rocks on all the islands of the St. Kilda group, where the species breeds in great numbers. The bird has also established itself in the Shetland Islands within the last quarter of a century, and on the Island of Handa and the mainland of Scotland, near to Thurso, quite recently.

*Materials.*—Dried grass and tufts of sea pink,

EGG OF FULMAR PETREL

generally nothing at all except a few chippings of stone.

*Egg.*—One; white and rough when newly laid, but quickly becoming soiled. Average size about 2.9 by 1.98 in.

*Time.*—May and June.

*Remarks.*—Resident, but wandering. Note: Seebohm says it is "a very silent bird," and Macgillivray, "I never observed them" (the birds at St. Kilda) "utter any cry when flying, or even when their nests were being robbed." Local and other names: Fulmar, Northern Fulmar, Mallemock, Mallduck, Malmock. Gregarious, and a close sitter.

HOME OF THE FULMAR PETREL, THE DOON, ST. KILDA.

---

## PETREL, LEACH'S FORK-TAILED.

(*Oceanodroma leucorrhoa.*)

Order TUBINARES ; Family PROCELLARIIDÆ (PETRELS).

*Description of Parent Birds.*—Length about seven and a quarter inches. Bill of medium length, nearly straight, and black. Irides dark brown. Head, neck, and back brownish-black, the two former rather lighter in shade than the latter. Wing-coverts reddish-brown, tinged with grey on the edges ; quills black. Upper tail-coverts white ; tail-quills black and slightly forked ; breast, belly, vent, and under tail-coverts, in the middle, black. A white streak starts from behind each thigh and runs down the sides of the vent and under tail-coverts. Legs, toes, webs, and claws black.

The female is like the male.

LEACH'S FORK-TAILED PETREL'S NEST AND EGG.

*Situation and Locality.*—In burrows made in soft peat earth, under rocks, holes, fissures, and clefts in rocks, and in holes of walls, close to the sea; on rocky islands, such as the St. Kilda group, Hebrides, and some of those off the Irish coast.

*Materials.*—Dry grass and bits of moss, sometimes nothing whatever.

*Egg.*—One; white, chalky, and speckled round the larger end with small rust-coloured and greyish-brown spots. Size about 1.3 by .96 in. (*See* Plate XIV.)

*Time.*—June.

*Remarks.*—Resident, but wandering. Notes: *pewr-wit, pewr-wit,* said to be uttered by the birds as they sit on their nests, both by night and day. However, whilst at St. Kilda I never once heard it, although I examined a goodly number of nesting burrows. Local and other names: Leach's Petrel, Fork-tailed Storm Petrel, Fork-tailed Petrel. Gregarious, and a very close sitter. I have on several occasions at St. Kilda taken the bird off her egg without protest on her part.

---

## PETREL, STORM.

(*Procellaria pelagica.*)

Order TUBINARES; Family PROCELLARIIDÆ (PETRELS).

*Description of Parent Birds.*—Length about six inches. Bill moderately long, hooked at the tip, and black. Irides dark brown. Head, neck, back, wings, and tail a uniform sooty-black. The outer edges of some of the smaller feathers of the wings and upper tail-coverts white. All the under-parts are sooty-brown, with exception of the sides of

SITUATION OF STORM PETREL'S NEST.

+ ENTRANCE HOLE.

the vent, which are white. Legs, toes, and webs black.

The female does not differ from the male.

*Situation and Locality.*—In old Puffin and rabbit-burrows, holes in cliffs, under large boulders, and in holes in walls. In the Scilly Islands, Lundy, at suitable places along the Welsh coast, the western and northern coasts of Scotland, and the islands lying off them; round the Irish coast, but neither on the east coast of England nor Scotland. Our illustration represents a boulder of rock under which a Stormy Petrel had its nest on Ailsa Craig for several years, according to an informant who had lived for a long time upon this sea fowl haunted rock, and who took a great interest in the doings of his feathered neighbours.

*Materials.*—A few blades of dry grass generally, but the egg is often laid on the bare ground.

*Egg.*—One. White, rough, and chalky in appearance, with small, dust-like, reddish-brown spots in an almost indistinguishable zone round the larger end. Size about 1.1 by .83 in. (*See* Plate XIV.)

*Time.*—End of May, June, July, and even as late as September.

*Remarks.*—Resident, but keeping to the open sea, except during the breeding season, or when driven ashore by violent storms. Notes: a warbling chatter, sung whilst the bird is sitting on her egg. Local and other names: Mother Carey's Chicken, Stormy Petrel, Little Petrel, Witch, Allamotti, Sea Swallow, Spency, Assilag, Mitty. Sits very closely.

WOODCOCK.
(See p. 498.)

DUNLIN.

COMMON SNIPE.
(See p. 570.)

COMMON CURLEW.

*Note.—In referring to the eggs the above names should be read from left to right.*

WOODCOCK.
(*See p.* 498.)

DUNLIN.
(*See p.* 102.)

COMMON CURLEW.
(*See p.* 62.)

WHIM
(*See p.*

NOTE.—*In referring to the eggs the above names should be read fro n*

PLATE 9.

## PHALAROPE, RED-NECKED.

(*Phalaropus hyperboreus.*)

Order Limicolæ ; Family Scolopacidæ (SNIPES).

PHALAROPE.

*Description of Parent Birds.*—Length from seven to eight inches. Bill of medium length, straight, slender, and black. Irides dark brown. Head, hind part of neck, back, wing-coverts, scapulars, and tertials dark ash-grey, the last two sets of feathers being tipped with rust colour. Wing-quills dusky, some of them being tipped with white. Rump and upper tail-coverts dusky, banded with white; tail-quills dusky or brownish-grey, darkest in the centre. Chin white; front and sides of neck rusty-red; upper breast grey, barred with white; under-parts white. Some specimens are white from chin to vent. Legs, toes, and membrane down either side of toes, green; claws black.

The female is rather larger, more richly coloured, and is easily distinguished from the male by reason of the fact that she has a white spot over the eye, whereas her mate is ornamented in the same place with a white streak. These distinctions, although clearly seen in our tailpiece, are rarely, if ever, shown in drawings of this species.

*Situation and Locality.*—On the ground, in tufts of grass, in a hollow on the top of a small hillock; on moors and mountains not far from the edge of a loch or pool, in the Orkneys and Hebrides.

I have devoted a considerable amount of time

RED-NECKED PHALAROPE'S NEST AND EGGS.

to the study of this species, and have always been charmed with its lively, confidential manner.

*Materials.*—Fine dead grass. The nest is deep, and about the size of that of a Titlark, but not so tidy in construction.

*Eggs.*—Four, ground-colour varying from olive-green to light buffish-brown, spotted and blotched with umber and blackish-brown, most thickly at the larger end. When the nest has been situated close to a muddy pool I have known the eggs so besmeared that the markings on them were almost invisible. Size about 1.1 by .83 in. Very pyriform and easily distinguished by small size. (*See* Plate VIII.)

*Time.*—June.

*Remarks.*—Migratory, arriving in May and departing in August. Note: *tirrr.* Local and other names: Red Phalarope, Half Web, Red-necked Coot Foot, Red-necked Lobe Foot. Gregarious, and very tame on its breeding grounds.

MALE AND FEMALE RED-NECKED PHALAROPES.

## PHEASANT.

(*Phasianus colchicus.*)

Order GALLINÆ; Family PHASIANIDÆ (PHEASANTS).

PHEASANT ON NEST.

*Description of Parent Birds.* — Length about three feet, nearly two of which are accounted for by the abnormally long tail. Beak short, curved downwards, and light yellowish horn colour, duller at the base. Irides hazel. Round the eye the skin is bare, crimson, and minutely speckled with black. The feathers of the head and neck all round are steel-blue, with a purple-green or brown sheen, according to the light upon them. Upper part of back deep orange, tipped with rich black; lower back and scapulars made up of a mixture of dark orange and dull brown, each feather having a straw-coloured outer margin. Wing-coverts red, of different shades. Quills greyish and yellowish-brown. Rump and upper tail-coverts pale yellowish-brown. Tail-quills brown, tinged with yellow, and transversely barred with black. Breast and belly golden red, each feather edged with rich black, glossed with purple and gold; vent and under tail-coverts dusky-brown. Legs, toes, claws, and spurs brownish-grey.

The female is about a foot shorter, and much subdued in coloration. Her plumage is composed principally of yellowish and dark brown.

PHEASANT'S NEST AND EGGS.

*Situation and Locality.*—On the ground, amongst coarse, long grass, in or near hedgerow bottoms, under bramble bushes, brackens, weeds, and scrub, on the outskirts of woods, plantations, and coppices all over the country where there is plenty of wood, water, and protection. Specimens have been found occupying a deserted squirrel's drey, in a Scotch fir, and on the tops of stacks. I have seen them out in bare, open fields on one or two occasions.

*Materials.*—A few dead leaves, dried grass-blades, bracken, or fern-fronds.

*Eggs.*—Eight to thirteen; as many, however, as seventeen have been found in one nest under circumstances which pointed to their having been laid by one hen, and in other cases even as many as twenty-six, undoubtedly the joint production of two hens, as the bird has often been known to share a nest, not only so far as laying was concerned, but sitting with other hens of its own species; also with the Partridge, and has been known to lay in the nests of the Red Grouse, Capercaillie, Wood Cock, and Wild Duck. Olive-brown is the general colour of the eggs, but specimens may be met with of a greyish-white, tinged with green or bluish-green. They are unspotted, but finely pitted. Size about 1.87 by 1.4 in. (*See* Plate XV.)

*Time.*—April and May, sometimes as early as March, and as late as September or October.

*Remarks.*—Naturalised, and holds its own only by protection. There can be but little doubt that the bird was first introduced by the Romans. Notes: crow of male, a short, loud cackle, and the note of the female, a shrill, piping whistle. Local or other name: none. Sits closely and, curiously enough, emits little or no scent at this period. In a wild state the bird is monogamous, but in this country, in its semi-domesticated condition, it is polygamous.

YOUNG PHEASANTS FEEDING.

---

## PINTAIL DUCK. *See* DUCK, PINTAIL.

*See* DUCK, PINTAIL.

---

## PIPIT, MEADOW.

(*Anthus pratensis.*)

Order PASSERES; Family MOTACILLIDÆ (WAGTAILS).

MEADOW PIPIT.

*Description of Parent Birds.*—Length about six inches. Bill of medium length, slender, straight, and dark brown, except at the base of the under-mandible, where it is of a lighter tint. Irides hazel. Crown, nape, back, and upper tail-coverts dark brown, the border of each feather being of a lighter greyish tint.

Wings brownish-black, the feathers being edged with light brown; tail dark brown, the two outer feathers on either side margined with white, the rest with light brown; chin and throat dull white; sides of neck and breast pale buffish-white or yellowish-brown, with numerous elongated dusky spots; belly and under tail-coverts dull white, tinged with brown. Legs and toes light brown; claws dusky, hind one long and curved.

The female is said to be slightly smaller, though the difference is not at all apparent; her plumage is similar.

*Situation and Locality.*—On the ground in the shelter of a tuft of grass, heather, bit of overhanging bank or stone. I have on several occasions seen nests in holes in banks and amongst rocks. The bird breeds commonly throughout the British Isles, but most numerously in pasture land and moorland districts, where it is more often victimised by the Cuckoo than any other British bird.

*Materials.*—Bents, bits of fine dead grass and horsehair.

*Eggs.*—Four to six, generally five; French grey, sometimes tinged with pale bluish-green, thickly covered with light or dusky-brown. The markings are generally so thickly distributed as to hide the ground-colour; indeed, I have met with specimens where none of it could be seen. Occasionally eggs may be found marked with hair-lines of dusky-black at the larger end. Size about .8 by .58 in. Distinguished by small size and brown appearance. (*See* Plate II.)

*Time.*—April, May, June, and occasionally as late as July.

MEADOW PIPIT'S NEST AND EGGS.

*Remarks.*—Migratory and resident, the latter being subject to local movement. Notes: song, short, soft, and musical; alarm notes, *trit, trit;* call, *zeeah, zeeah, zeeah.* Local and other names: Titling, Moor Tite, Titlark, Ling Bird, Teetick, Moss Cheeper, Wekeen, Pipit Lark, Heather Lintie, Moor Titling, Moor Tit, Meadow Lark. Sits closely, and hovers round, uttering its note, *trit, trit.*

YOUNG MEADOW PIPITS.

---

## PIPIT, ROCK.

(*Anthus obscurus.*)

Order PASSERES; Family MOTACILLIDÆ (WAGTAILS).

*Description of Parent Birds.*—Length about six and three-quarter inches. Bill medium, nearly straight, slender, and dark brown, except at the base, where it is dull orange. Irides dark brown. A faint yellowish-white streak runs over the eye and ear-coverts. Crown, nape, back, rump, upper tail-coverts, and wings dull brown, slightly tinged with green, the feathers being streaked along the centre with a darker shade. Edges and tips of wing-quills lighter. Tail dark brown; outer feathers

ROCK PIPIT'S NEST AND EGGS.

on either side greyish, on the exterior webs and at the tips. Chin and throat greyish, or dull yellowish-white, the latter and sides of neck mottled and streaked with brown. Breast dull greenish-white, streaked and spotted with brown. Sides olive-brown; belly, vent, and under tail-coverts dull yellowish-white, sparingly streaked with brown. Legs and toes reddish-brown; claws black.

The female is a trifle smaller, but similar in plumage.

*Situation and Locality.*—On ledges and in crevices of rock; under an overhanging piece of stone, or in the shelter of a tuft of grass growing on rocky sea coasts, pretty generally round our shores, with the exception of Essex, Suffolk, Norfolk, and Lincoln.

*Materials.*—Seaweed and dry grass of various kinds, with an inner lining of finer grass, and occasionally horsehair.

*Eggs.*—Four or five, grey in ground-colour, slightly tinged with green or reddish-brown, minutely and closely spotted and mottled with underlying markings of grey, and surface spots of reddish-brown, occasionally marked at the larger end with one or two dark brown lines. The spots are, as a rule, more numerous at the larger end. Size about .85 by .63 in. Distinguished by large size and locality of the nest. (*See* Plate II.)

*Time.*—April, May, June, and July.

*Remarks.*—Resident, although numbers migrate. Notes: call, a shrill *hist* or *pst*. Local and other names: Shore Pipit, Rock Lark, Sea Titling, Dusky Lark, Field Lark, Sea Lintie. Sits closely.

## PIPIT, TREE.

(*Anthus trivialis.*)

Order Passeres; Family Motacillidæ (WAGTAILS).

TREE PIPIT.

*Description of Parent Birds.*—Length about six and a half inches. Bill of medium length, nearly straight, slender, and dark brown, lighter on the edges and at the base. Irides hazel. Crown, nape, and back dark brown, the feathers being bordered with lighter brown; wings darkish-brown; lesser coverts edged and tipped with greyish-white; greater coverts edged with pale brown; these two lighter colours form distinct bars across the wings. Rump, upper tail-coverts, and quills brown, the two outer ones on each side nearly all dirty white; chin and throat pale brownish-white. A brown streak runs from the gape, slightly backward, and for some distance downwards. Sides of neck and breast pale buff, with streaks of brown on the former and round spots on the latter; belly, vent, and under tail-coverts dirty white. Legs, toes, and claws pale yellowish-brown.

The female is a little smaller in size, and the spots on her breast are not so large.

This species is larger than the Meadow Pipit, has a stronger beak, fewer and larger spots on the breast, and the claw on the hind toe is shorter.

*Situation and Locality.*—On the ground, concealed by a tuft of grass, in hedgerow banks, on the sloping banks of streams, hidden by a low weed-tangled bush. The bird seems fond of the same locality, and I know several places in Yorkshire where pairs return to nest year after year with the utmost regularity; the cocks using the same tree, often the very same branch, to start from and return to after their short singing flight. Near woods, plantations, and tree-fringed streams. Affects more cultivated districts than the Meadow Pipit. Scattered over England in suitable districts, rare in the west and parts of Wales, more numerous in the south of Scotland, rare in the north, and not reported in Ireland on trustworthy authority. Our illustrations were procured in Surrey.

*Materials.*—Dry grass, moss, roots, lined with finer grass and generally, though not always, horsehair.

*Eggs.*—Four or five, sometimes six; exceedingly variable in coloration. Professor Newton regards those of "a french-white, so closely mottled or speckled with deep brown as almost to hide the ground-colour," as the normal type; whilst Morris regards those "greyish-white in ground-colour, with a faint tinge of purple, clouded and spotted with purple-brown or purple-red," as the most general, and amongst the nests I have found this type has certainly been the most numerous. In another type the ground-colour is yellowish-white, and the spots rich reddish-brown. Some eggs are of a uniform brownish-pink, rarely marked on the larger end with hair-lines of dark brown or black. The spots not only vary much in colour, but in size and distribution. Size about .83 by .63 in. The larger size of the eggs, their inclination to reddish-brown, and locality of nest, help to distinguish them. (*See* Plate III.)

*Time.*—May and June.

*Remarks.*—Migratory, arriving in April and

TREE PIPIT'S NEST AND EGGS

departing in September or October. Notes: a sweet, ringing *tsee, tsee, tsee*, uttered pretty quickly. Local and other names: Field Lark, Field Titling, Pipit Lark, Tree Lark, Grasshopper Lark. Sits closely, especially when incubation is advanced.

---

## PLOVER, GOLDEN.

(*Charadrius pluvialis.*)

Order LIMICOLÆ; Family CHARADRIIDÆ (PLOVERS).

YOUNG GOLDEN PLOVER.

*Description of Parent Birds.*—Length between eleven and twelve inches. Bill of medium length, straight, and black. Irides brownish-black. On the forehead is a band of white; crown, nape, neck, back, wing-coverts, rump, and tail-coverts, deep greyish-brown, the edges and tips of the feather being marked with yellow; wing-quills brownish-black; tail dark brown, barred with brownish-black and greyish-white. Chin, throat, sides of neck, breast, and belly deep rich black, bordered on the sides below the wings by a band of white; under tail-coverts white. Legs, toes, and claws black.

The female resembles the male, but both are subject to some little variation, depending upon constitutional vigour.

*Situation and Locality.*—On the ground. I have met with nests in short closely-cropped heather

GOLDEN PLOVER'S NEST AND EGGS.

that were nearly as cup-shaped as the nest of the Chaffinch, and in which the four eggs were almost standing on their sharp ends. Amongst fringe moss, coarse grass, and short heather in the rough moorland and wild boggy parts of Somerset and Devon, the north of England, Wales, Scotland, and Ireland.

*Materials.*—A few pieces of dry grass, rushes, or heather-tops, forming a lining to the hollow of the nest.

*Eggs.*—Four, pear-shaped, yellowish stone or cream colour, blotched and spotted with umber-brown and blackish-brown. They are larger than those of the Lapwing, not quite so pyriform, and lack the olive tinge in their ground-colour. The birds also nest, as a rule, on higher and wilder ground. Size about 2.07 by 1.4 in. (*See* Plate VII.)

*Time.*—May and June.

*Remarks.*—Resident, but subject to local and partial migration. Notes: *tlui*, and *taludl*, *taludl*, *taludl*, the first note being uttered with low and melancholy deliberation and the latter hurriedly. Local and other names: Yellow Plover, Whistling Plover. Does not sit very closely as a rule; however, I have known the bird do so before incubation was far advanced, and feign a broken wing in order to decoy the intruder away.

GOLDEN PLOVER ON NEST.

## PLOVER, GREAT. *See* CURLEW, STONE.

---

## PLOVER, KENTISH.

(*Ægialitis cantiana.*)

Order LIMICOLÆ ; Family CHARADRIIDÆ (PLOVERS).

*Description of Parent Birds.*—Length about seven inches. Bill shortish, nearly straight, and black. Irides brown. Forehead and a line running over the eye and ear-coverts white ; middle crown black ; back of head yellowish-brown. A black streak commences at the base of the beak, and passing through the eye, includes the ear-coverts ; nape white ; back, wings, and upper tail-coverts ash-brown, with exception of the wing-primaries, which are dull black, edged with white on some of the outside shafts ; tail-quills ash-brown towards the base, dusky-black towards the tip, and white on the outsides ; chin, cheeks, and sides of upper part of neck white ; sides of lower part of neck, just in front of the shoulder or point of the wing, black ; breast, belly, and under-parts white. Legs, toes, and claws dark slate-colour.

The female differs only in having the black on the head and sides of lower neck less distinct and covering a smaller area. This species may be distinguished from Ringed Plover by smaller size and lighter colours.

*Situation and Locality.*—In a hollow of the sand or shingle ; sometimes on dry seaweed which has been cast up by the waves. The breeding area of the bird is very limited indeed, and its numbers small. In suitable places along the coast between Hastings and Dover.

KENTISH PLOVER'S EGGS.

*Materials.*—None, the eggs being deposited in a slight hollow.

*Eggs.*—Three or four, generally the former number; cream, stone, or dark buff in ground-colour, streaked, spotted, and blotched with brownish-black and very dark grey. More pyriform than those of the Lesser Tern, and distinguished from those of the Ringed Plover by scrawl-like character of markings. Size about 1.25 by .9 in. (*See* Plate VII.)

*Time.*—May.

*Remarks.*—Migratory, arriving in April or early in May, and departing in August or the beginning of September. Notes: call, *tirr, tirr, pitt, pitt, pwee, pwee;* alarm, a plaintive and also a sharp whistle. Local or other name: none. In Yarrell's time the eggs were in great demand as table delicacies, and dogs were trained to find them. It is very satisfactory to be able to state that the bird and its eggs are now enjoying the personal protection of a paid watcher. Does not sit closely, and runs on quitting the eggs.

KENTISH PLOVER ON EGGS.

## PLOVER, NORFOLK. *See* CURLEW, STONE.

---

## PLOVER, RINGED. *Also* RINGED DOTTEREL.

(*Ægialitis hiaticula.*)

Order LIMICOLÆ; Family CHARADRIIDÆ (PLOVERS).

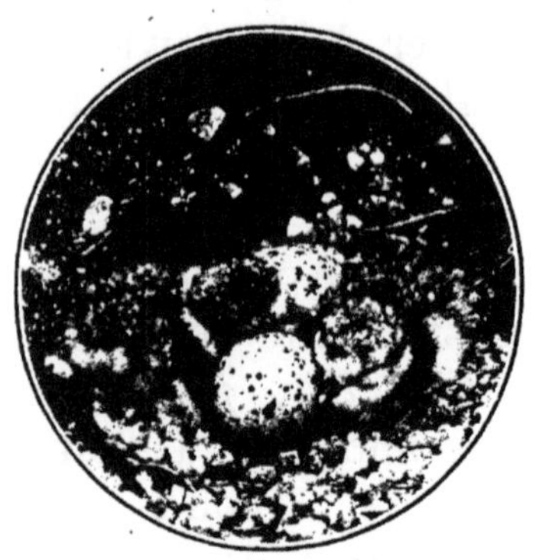

YOUNG RINGED PLOVERS AND EGGS.

*Description of Parent Birds.*—Length about seven and three-quarter inches. Beak short, straight, black at the tip, and rich, dark yellow towards the base. Irides brown. Forehead white, middle of crown black, followed by greyish-brown, which extends down the back of the neck; back and wings greyish-brown, except the ends of the coverts, which are tipped with white, and the primaries, which are dusky, with some white at the base and along the shafts. Upper tail-coverts greyish-brown; quills greyish-black in the centre, outside feathers white. A black patch commences at the gape and passes under the eye, backward and downward, to the side of the neck. A broadish collar of white passes round the upper part of the neck, followed by a gorget of black, which is deepest in front. Breast and all the under-parts white. Legs and toes orange; claws black.

In the female the black parts of the head and neck are not so broad or well defined.

*Situation and Locality.*—On the plain surface or in a slight hollow, scraped in the sand or shingle above high-water mark on stretches of flat, sandy

RINGED PLOVER'S NEST AND EGGS.

shores, also in shallow crevices of bare, flat sea-shore rocks; sometimes quite inland on the banks of rivers and lakes in nearly all suitable places throughout the British Isles.

*Materials.*—None generally, but sometimes a lining of small pebbles; and in places where a crevice in a flat rock has been adopted I have often met with a lining of small sea shells.

*Eggs.*—Four, pear-shaped, pale buff, cream, or stone colour, spotted with smallish and pretty evenly distributed dots of black, blackish-brown, and bluish-grey, which distinguish them from those of the Kentish Plover. Size about 1.4 by 1.0 in. (*See* Plate VII.)

*Time.*—April, May, and June, although eggs have been found as early as March and as late as August.

*Remarks.*—Resident, but subject to much local movement. Notes: alarm, *pen-y-et.* Local and other names: Sand Lark, Dull Willy, Sand Laverock, Ringed Dotterel, Stonehatch. Does not sit closely. Eggs harmonise very closely with their surroundings.

RINGED PLOVER SITTING ON EGGS.

## POCHARD.

(*Fuligula ferina.*)

Order ANSERES ; Family ANATIDÆ (DUCKS).

*Description of Parent Birds.*—Length about nineteen inches. Bill of medium length, depressed near the centre, black at the tip, pale blue in the middle, and black at the base, which is slightly raised. Irides red. Head and upper part of neck, all round, deep chestnut, lower part of neck, all round, black. Back, scapulars, tertials, and wing-coverts finely freckled with grey and dusky undulating lines ; secondaries and primaries bluish-grey, the former tipped with white and the latter ending in dusky-brown. Rump and upper tail-coverts dusky-black ; tail feathers dusky-brown mixed with grey. Breast, sides, and under-parts greyish-white, marked by minute dusky lines, darkest on the vent. Legs, toes, and webs leaden grey.

The female differs considerably in appearance. Her bill is black. Irides reddish-brown. Head and upper part of neck dull greyish-brown, lightest in front ; lower portion of neck dusky-brown. Upper-parts darker, and under-parts dull greyish-white, clouded with brown ; under tail-coverts dusky grey.

*Situation and Locality.*—On the ground, in tufts of rushes, coarse grass, in osier beds ; amongst flags and sedges growing on the shores of lakes, broads, and tarns in the north, east, and south of England ; also in Wales, Scotland and Ireland. Our illustrations were procured in Norfolk. The nest was situated amongst the reeds on the right-hand side of the picture on page 301.

*Materials.*—Sedges, rushes, and dry grass, with

POCHARD'S NEST AND EGGS.

an inner lining of down. The tufts are brownish-grey with whitish centres.

*Eggs.*—Seven to ten, occasionally as many as thirteen or fourteen. Pale greyish-buff or greenish-drab. Distinguished from the Tufted Duck only by the browner colour of the down tufts. Size about 2.35 by 1.7 in.

*Time.*—May.

*Remarks.*—Resident and migratory, being more numerous in winter. Notes: a low whistle, but when alarmed or vexed, a hoarse kind of croak, like *kr kr-kr.* The Pochard has been induced to breed season by season in some parts of the country by turning a pair of pinioned birds on to a suitable sheet of water, and thus accustoming their descendants to look upon the place as their natural breeding haunt. Local and other names: Red-headed Wigeon, Duncur, Red-headed Poker, Dunbird, Vare-headed Wigeon, Attile Duck, Blue Poker, Great-headed Wigeon. Sits closely.

NORFOLK BREEDING HAUNT OF THE POCHARD.

## PTARMIGAN.

(*Lagopus mutus.*)

Order GALLINÆ ; Family TETRAONIDÆ (GROUSE).

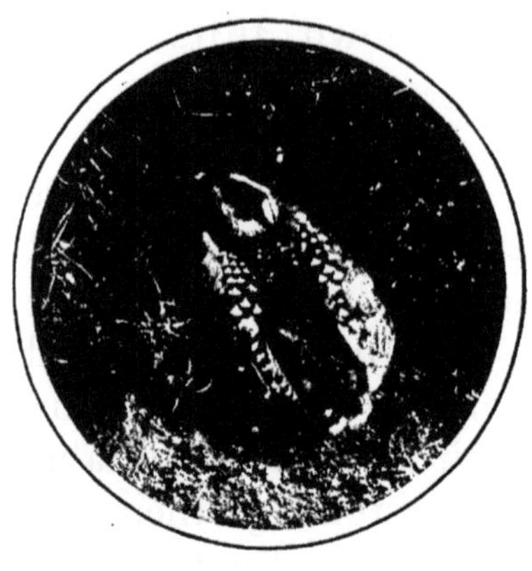

YOUNG PTARMIGAN HIDING.

*Description of Parent Birds.* — Length fifteen inches. Bill short, strong, curved downwards, and black. Irides hazel. Over the eye is a piece of erectile skin of a bright red colour. Head and neck barred and mottled with black, rusty-brown, and white or grey; back and upper tail-coverts pale brown or ash, mottled with small dusky spots and bars. Wings white, the shafts of the quills being black. Tail-quills black, tipped with white, the two centre feathers sometimes grey; chin white; throat white, mottled with brown; breast same as back; belly, vent, and under tail-coverts white. Legs and feet dull white; claws black.

The female is a trifle smaller; her head and upper-parts have more red, rusty-yellow, and black, and less grey than in the case of the male. The dark parts on the wing-quills are broader, and her under-parts are darker.

*Situation and Locality.*—On the ground, amongst heather and the short stunted vegetation growing on the rock-strewn and bleak mountains of the Highlands of Scotland, and some of the larger islands of the Hebrides.

*Materials.*—A few bits of dead heather, dry

PTARMIGAN'S NEST AND EGGS.

PTARMIGAN ON NEST.

grass, or leaves, used as a lining to the hollow chosen for the reception of the eggs.

*Eggs.*—Seven to ten or twelve, greyish-white to pale red-brown in ground-colour, blotched and spotted all over with very dark, rich brown. Distinguished from the eggs of the Red Grouse by their buffy ground-colour and smaller number of markings. Size about 1.7 by 1.1 in. (*See* Plate XV.)

*Time.*—May and June.

*Remarks.*—Resident. Notes: sometimes low, and at others a kind of loud and prolonged croak. Local and other names: Rock Grouse, White Grouse, White Partridge (from the fact that the bird turns white in winter), White Game. Sits very closely. The nest is somewhat difficult to find. I have seen one under the shelter of a crag jutting from a steep hillside.

BREEDING HAUNT OF THE PTARMIGAN.

*Photographed close upon M.dsummer.*

## PUFFIN.

(*Fratercula arctica.*)

Order PYGOPODES ; Family ALCIDÆ (AUKS).

PUFFIN.

*Description of Parent Birds.*—Length about twelve inches. Bill rather short, and deeper than it is long. Both mandibles are arched from base to tip, the upper one being a trifle hooked. It is of such clumsy appearance as to suggest a kind of sheath over the real bill. It is much compressed sideways, and furrowed transversely. The basal ridge is yellow; then occurs a space of bluish-grey, followed by four ridges, and three grooves of a rich orange colour. There is a space of naked skin at the gape, which is yellow. Irides grey. Cheeks and ear-coverts dirty white; forehead, crown, back of head, ring round neck, back, wings, and tail, black. Breast, belly, and vent white. Legs, toes, and webs orange; claws black.

The female has a slightly narrower bill.

*Situation and Locality.*—In burrows of varying length, dug by the bird's own exertions, in peat or mould, or taken from a rabbit by force; sometimes amongst fallen rocks or in crevices of cliffs. Our full-page illustration is from a photograph taken on the Farne Islands, where a large colony nests yearly. I took the egg from the end of one of two burrows having the same entrance, and placed it in front so as to show in the picture. In walking across the top of the island, which is covered by a soft layer of peat, the visitor feels the earth giving way beneath his feet at each step, so much is it honeycombed by the birds, which scuttle out of their burrows in all directions. Breeds at the Farne Islands, Flamborough Head, parts of the south coast, Scilly Isles, west coast of England, Wales, Lundy, west coast of Scotland, Hebrides, St. Kilda in vast numbers, the north of Scotland, and in suitable places round the Irish coast.

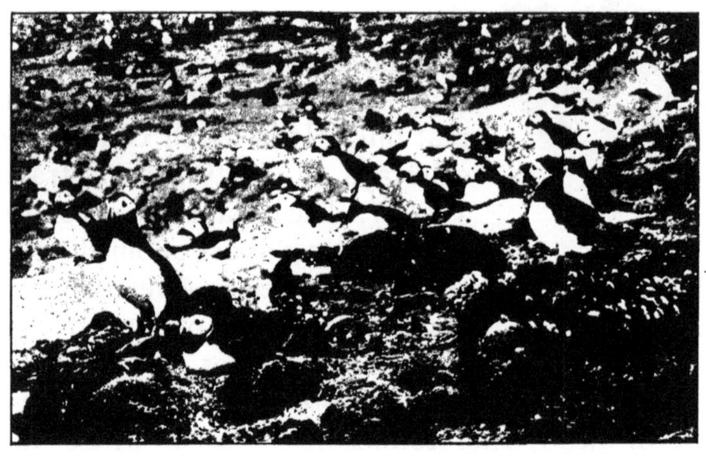

PUFFINS AT HOME.

*Materials.*—Occasionally a few bits of grass, feathers, or roots; often nothing whatever.

*Egg.*—One; dull white or grey, marked with a few indistinctly defined spots of pale brown and grey, generally at the larger end. They soon become soiled and dirty from contact with their surroundings. Size about 2.4 by 1.67 in. (*See* Plate XIII.)

*Time.*—May and June.

*Remarks.*—Migratory, arriving in April and departing in August. More northern breeding birds winter with us. Notes: *o-r-r* to *a-r-r*, according to

PUFFIN'S NESTING BURROW AND EGG.

the bird's state of mind. Local and other names: Tammy Norrie, Coulter-Neb, Sea Parrot, Tommy Tomnoddy, Ailsa Cock, Cockandy, Lunda Bonger, Gulderhead, Bottlenose, Pope, Marrot, Mullet. Sits closely.

---

## QUAIL.

(*Coturnix communis.*)

Order GALLINÆ; Family PHASIANIDÆ (PHEASANTS).

*Description of Parent Birds.*—Length about seven inches. Beak rather short, strong, slightly curved, and dusky. Irides hazel. A broadish, pale wood-brown streak commences at the base of the beak, and passes over the eye and ear-coverts. The crown is also divided by a much narrower streak of the same colour. The feathers of the head, neck, back, rump, and tail are brown, with lighter coloured shafts and longitudinal streaks of wood-brown. Wing-quills dusky-brown, with small rust-coloured bands on the outer webs. Chin and throat white, crossed by two dark brown gorgets; breast pale rusty-brown with lighter shafts; under-parts yellowish-white; flank-feathers pale buff in the centre, mottled and barred on the margins with brown. Legs, toes, and claws pale brown.

The female lacks the gorgets, and has the breast feathers marked with dark spots on either side of the pale shafts.

*Situation and Locality.*—On the ground in growing cornfields, grass and clover fields, very sparingly throughout the British Isles.

*Materials.*—The slight hollow used as a nest is scantily lined with blades of grass, trodden down, or a few dead leaves.

HERRING GULL.
(*See p.* 169.)

BLACK-HEADED GULL.
(*See p.* 160.)

KITTIWAKE.
(*See p.* 206.)

GREAT BLACK-BACKED GULL.
(*See p.* 166.)

NOTE.—*In referring to the eggs the above names should be read from left*

*Eggs.*—Seven to twelve ; as many even as twenty have been found, doubtless the production of two hens. Pale yellowish-brown of varying shades, spotted, blotched, and clouded with umber-brown and blackish-brown. Some eggs are only spotted, whilst others are thickly clouded with varying shades of brown. Distinguished by small size and character of markings. Size about 1.1 by .9 in. (*See* Plate XV.)

*Time.*—May and June.

*Remarks.*—Migratory and resident. Notes : *click-clic-lic.* Other names : none. Sits closely.

---

## RAIL, LAND. *See* CRAKE, CORN.

---

## RAIL, WATER.

(*Rallus aquaticus.*)

Order FULICARIÆ ; Family RALLIDÆ (RAILS).

WATER RAIL AND NEST.

*Description of Parent Birds.*—Length nearly twelve inches. Beak rather long, nearly straight, and red. Irides hazel. Crown, neck, back, and wing-coverts olive- or reddish-brown, with a deep black mark in the middle of each feather ; wing and tail primaries dusky, with lighter margins ; chin, cheeks, throat, sides of neck, breast, and belly, leaden-grey ; sides and flanks deep slaty-grey, with bars of white ; vent buffish ; under tail-coverts greyish-white. Legs and toes reddish-brown.

The female resembles the male, although her

beak is not so long or her plumage so bright and distinctive; she also generally shows some white bars, which the male lacks, on the wing-coverts.

*Situation and Locality.*—On the ground amongst long grass, a clump of rushes or reeds, in thick osier beds, swamps where alders grow, round ponds and ditches, on the banks of slow-running rivers and in boggy ground abounding in reeds and dense aquatic growths. Said to nest generally throughout the United Kingdom, but nowhere abundant. I have only met with it breeding in East Anglia.

WATER RAIL SITTING.

*Materials.*—Reeds, sedge grass, and flags, in rather liberal quantities.

*Eggs.*—Five to eleven, generally six or seven; creamy-white in ground-colour, speckled with a few small reddish spots and underlying dots of ash-grey. The locality of the nest and small spots distinguish them from those of the Corn Crake. Size about 1.4 by 1.0 in. (*See* Plate XV.)

*Time.*—April, May, June, and July.

*Remarks.*—Migratory and resident. Many of our winter visitors retire north to breed. Notes: a

WATER RAIL'S NEST AND EGGS.

soft *whit*, heard after dusk ; also a harsher, grating kind of note. Local and other names : Runner, Skiddycock, Brook-runner, Bilcock, Velvet-runner, Grey-skit, Oarcock. A pretty close sitter, slipping away without demonstration.

---

## RAVEN.

(*Corvus corax.*)

Order PASSERES ; Family CORVIDÆ (CROWS).

ADULT RAVEN AT HOME.

*Description of Parent Birds.*—Length about twenty-six inches. Beak of medium length, curved downward towards the tip, stout, and black. Irides brown and grey. At the base of the beak are a number of coarse hairs pointing forward. The plumage is a uniform black, glossed with a purple and blue sheen. Legs, toes, and claws black. The female is a little smaller, and less glossed.

*Situation and Locality.*—In crevices and on ledges of high inaccessible cliffs, either on the seashore or inland. The bird sometimes nests in tall trees, and is to be found in the wild unfrequented parts of England, Wales, Scotland, and Ireland. It is interesting to note that it bred in Caterham Valley, not far from Croydon, some seventy years ago, and in Epping Forest until a much later period.

*Materials.*—Sticks of various sizes, heather

RAVEN'S NEST.

stalks, or dry seaweed, with an inner lining of grass, wool, and hair.

*Eggs.*—Five to seven; greyish-green, bluish-green, or greenish-brown, blotched, splashed, and spotted with dark greenish or smoky brown, and underlying markings of a lighter greyish-purple tinge. Variable both in regard to coloration and size, but generally distinguishable from those of the Carrion Crow and Rook by their larger size. Average measurement about 1.95 by 1.3 in. (*See* Plate I.)

*Time.*—February, March, and April.

*Remarks.*—Resident. Note: a deep hoarse *cronk*, that may be heard at great distances. Local and other names: Corbie, Corbie Crow, Great Corbie Crow. Sits lightly, and shows a wonderful devotion to old haunts. The central feathers in a Raven's tail are longer than the rest, and thus serve to identify the bird when it is sailing high overhead.

---

## RAZORBILL.

(*Alca torda.*)

Order PYGOPODES; Family ALCIDÆ (AUKS).

RAZORBILL.

*Description of Parent Birds.*—Length about seventeen or eighteen inches. Bill fairly long, straight, except towards the tip, where it is much decurved, and black. A white, curved line runs across both mandibles, and a well defined one from the top of the bill to the eyes. The basal half of the beak is covered with feathers. Irides dark brown. Crown, nape, back, wings, and tail black, with the exception of a narrow

RAZORBILL'S EGG.

band of white across the wings; chin and throat dark brown; breast and all under-parts snowy white. Legs, toes, and webs brownish-black; claws black.

Female similar.

*Situation and Locality.*—In crevices, crannies, under crags, and on ledges of high maritime cliffs; pretty generally round our coasts. One observer says that he has found its egg in a Puffin burrow, and another in a Cormorant's nest at the Farne Islands. The latter is a somewhat remarkable

RAZORBILLS AT HOME.

circumstance, inasmuch as the Cormorants exclusively occupy a rock there, and I have never once seen a Razorbill breeding on the Northumbrian coast. Our full-page illustration is from a photograph taken on Ailsa Craig, where great numbers breed.

*Materials.*—None; the egg, when laid on bare, flat rock, is often swept off by a gust of wind.

*Egg.*—One; varying from white to buffy-white, or even reddish-brown, spotted and blotched with large, bold, and numerous markings of greyish, chestnut, reddish and blackish brown. Not so pointed as that of the Guillemot, and interior of shell greenish instead of yellowish-white, which is the colour of all varieties of Guillemots' eggs, except those with intense green or blue ground-colours, when blown. Average size about 2.9 by 1·87 in. (*See* Plate XIV.)

*Time.*—May and June.

*Remarks.* — Resident, but wandering, except during the breeding season. Notes: a kind of grunting noise when disturbed. Local and other names: Black-billed Auk, Marrot, Murre, Razor-bill Auk, Sea Crow, Bawkie, Alk or Oke, Falk; in Zetland, Hiogga. Gregarious, and a close sitter.

---

## REDBREAST. *See* ROBIN.

---

## REDPOLL, LESSER.

(*Linota rufescens.*)

Order PASSERES; Family FRINGILLIDÆ (FINCHES).

*Description of Parent Birds.*—Length about four and a quarter inches. Bill short, nearly conical, thick at the base, and brownish horn colour. Irides dusky-brown. Crown, to beyond the line of the eyes, crimson red; sides of head brown. Hinder part of crown, nape, back, rump, and upper tail-coverts dark brown, the feathers being bordered with light reddish-brown, slightly mixed with grey; tail-coverts tinged with crimson. Wings dusky, the feathers being edged with pale brown; the middle and greater coverts are tipped with light reddish-brown, making two rather showy bars across. Tail slightly forked and dusky, edged outwardly with pale brown. Chin black, throat and breast rose-pink to vermilion; the middle of the breast, flanks, and under-parts light greyish. Sides streaked with dull brown. Legs and toes darkish brown; claws nearly black.

The female is a trifle smaller, and lacks the red on her breast and upper tail-coverts. Chin brownish-black; under-parts brownish-white, streaked with dark brown on the breast, sides, and flanks.

*Situation and Locality.*—In alders, willows, elms, firs, hawthorns, hazel and other trees and bushes, generally pretty low down, but sometimes at a considerable height. Occasionally it may be found in a heather tuft. In shrubberies, coppices, plantations, and bushes that fringe streams and ponds in mountain districts. It has been found breeding in nearly every county in England, but is most numerous in the northern counties, in Wales, and in Scotland. It breeds in Ireland most numerously in the north.

*Materials.*—Fine twigs (used as a foundation), dry grass, stalks, moss, and roots, with an inner lining of willow down, sometimes hair and feathers. It is cup-shaped, and, as a rule, a well made and beautiful little structure.

*Eggs.*—Four to six, generally five, of a very pale bluish-green colour, spotted generally about the larger end with orange-red, occasionally streaked with a darker colour. There are also underlying markings of pale greyish-brown. The black chin and smaller size of the parent birds distinguish the nest of this bird from those of either the Linnet or Twite. Size about .62 by .46 in. (*See* Plate II.)

*Time.*—May and June, as a rule, although nests containing fresh eggs may sometimes be found as late as the third week in July.

*Remarks.*—Resident in Scotland and the north

NESTS AND EGGS OF LESSER REDPOLL.

of England. A winter visitor further south, generally speaking, though specimens have stayed and bred. Notes: call, *peewit* and *kreck, kreck, hayid!* Song meagre but lively. Local and other names: Lesser Redpoll Linnet, Lesser Red-headed Finch, Rose Linnet. Sits very closely indeed.

---

## REDSHANK.

(*Totanus calidris.*)

Order LIMICOLÆ; Family SCOLOPACIDÆ (SNIPES).

YOUNG REDSHANK.

*Description of Parent Birds.*—Length about eleven iuches. Bill long, straight, slender, and dusky at the point and reddish at the base. Irides hazel. Crown, nape, back and wing-coverts greyish-brown, spotted and streaked with black; secondaries tipped with white; primaries nearly black. Rump, tail-coverts, and feathers white, the last barred with dusky black. Over the eye is a white streak, and from the gape to the eye a dusky-brown one. Chin, throat, breast, and under-surface of the body greyish-white, spotted and streaked with brownish-black. Legs and toes red; claws black.

The female resembles the male, but is larger.

*Situation and Locality.*—On the ground, in a little hollow or depression, sheltered by an overhanging tuft of coarse grass or heather, or in the crown of a rush-root, generally very well concealed;

REDSHANK'S NEST AND EGGS.

REDSHANK COVERING HER YOUNG.

in fen, marsh, and boggy districts on the swampy shores of mountain tarns and lochs; in the eastern and northern counties of England; in Wales and suitable parts of Scotland and Ireland.

*Materials.*—A few blades of grass or bits of moss; often nothing at all. The bird has a trick of scratching out several nesting places before finally selecting one for use.

*Eggs.*—Four and on very rare occasions five,

HAUNT OF REDSHANK.

much pointed at the smaller end; ground-colour varying from pale straw to buffish-brown, spotted and blotched with rich dark brown, and underlying markings of light brown and grey. Distinguished by the buff ground-colour and bold blotches. Size about 1.78 by 1.23 in. (*See* Plate IX.)

*Time.*—April, May, and June.

*Remarks.*—Resident, but subject to local movement. Notes: alarm, a shrill, discordant cry, resembling *took* or *tolk*. Local and other names: Redshank Sandpiper, Pool Snipe, Red-legged Horseman, Sandcock, Red-legged Sandpiper, Teuke. Sits lightly in most instances, although I have known individuals allow me to touch them on the nest. When incubation has advanced, the bird resorts to various alluring tricks to decoy the intruder away from her eggs and is exceedingly noisy. Young Redshanks can swim with ease and expedition.

---

## REDSTART.

(*Ruticilla phœnicurus.*)

Order PASSERES ; Family TURDIDÆ (THRUSHES).

MALE REDSTART.

*Description of Parent Birds.*—Length about five and a quarter inches. Bill of medium length, straight, and black. Irides hazel; forehead white; crown, nape, back, scapulars, and wing-coverts deep bluish-grey; wing-quills brown; rump, upper tail-coverts, and tail-quills bright rusty-red, with exception of the two middle feathers of the tail, which are brown like the wings. Over the base of the upper mandible, chin, cheeks, throat, and sides of the neck black. Breast bright rust colour, belly paler, vent yellowish-white, under the tail red. Legs, toes, and claws dark brown.

The female lacks the black and white on the head; has the upper-parts greyish-brown, tail duller, and under-parts fainter.

*Situation and Locality.*—In holes in trees, walls, thatches, rocks, and ruins; occasionally in such situations as inverted plant pots and disused pumps;

sparingly distributed over England, Wales, and Scotland, but rarely met with in Ireland. According to my experience, it is far more numerous in the dales in the north of England and in the Welsh valleys than anywhere else.

FEMALE REDSTART.

*Materials.*—Dead grass, rootlets, and leaves, with an inner lining of hair and feathers.

*Eggs.*—Four to six, occasionally eight, although I have never seen more than seven, of a pale, bluish-green, unspotted, and polished. Some text-books say, sometimes with a few faint red specks, but I have never found a clutch of eggs marked in any way. Situation of nest prevents confusion with Hedge Sparrow, and a sight of parent birds distinguishes eggs from those of Pied Flycatcher. All these three species frequently breed in the same locality. Size about .75 by .54 in. (*See* Plate IV.)

REDSTART'S NESTING PLACE.

*Time.* — May, June, and July.

REDSTART'S NEST AND EGGS.

*Remarks.*—Migratory, arriving in April and leaving in September. Notes: song, pleasing and imitative, uttered very late in the evening and very early in the morning while the female is sitting; call, represented by some authorities as *chippoo*, and others as *oichit*. Local and other names: Red-tail, Fire-tail, Bran-tail, Fire-flirt. Sits very closely indeed.

---

## RING OUZEL. *See* OUZEL, RING.

---

## ROBIN. *Also* REDBREAST.

(*Erithacus rubecula.*)

Order PASSERES; Family TURDIDÆ (THRUSHES).

YOUNG ROBIN.

*Description of Parent Birds.*—Length about five and three-quarter inches. Bill of medium length, nearly straight, and black; crown, nape, back, wings, and tail, olive-brown. Round the base of the beak, eyes, and upon the throat and upper breast, orange-red, succeeding which is a narrow space of bluish-grey; the rest of the under-parts white, tinged with brown on the sides, flanks, and under tail-coverts. Legs, toes, and claws reddish-brown.

The female is slightly smaller, and her coloration is not quite so bright.

*Situation and Locality.*—I have found the nest of the Robin in almost every conceivable situation—holes in banks, in walls inside and outside buildings, amongst ivy growing against walls and trees, in

ROBIN'S NEST AND EGGS.

flower pots, tea and coffee pots, old kettles, canisters, baskets, in rabbit holes, in the sides of hayricks, and other odd places. Common nearly everywhere throughout the British Isles.

*Materials.*—Fibrous roots and moss lined with dead leaves and hair.

*Eggs.*—Five or six, occasionally as many as seven and even eight; white or very light grey, blotched and freckled with dull light red. Sometimes the spots become confluent over nearly the entire surface of the shell, at others they are collected round the larger end. Occasionally very sparingly supplied or altogether absent. I have met with specimens almost as distinctively marked as those of a Great Tit. Size about .8 by .6 in. (*See* Plate IV.)

*Time.*—March, April, May, June, and July.

*Remarks.*—Resident and migratory. Some naturalists are of opinion that the Robins which inhabit our gardens and orchards in winter migrate North in summer, and that their places are supplied by more Southern members of the species. Anyway, it is certain that the bird does migrate, from the fact that specimens visit our lightships during the great autumn rushes. Notes: call, sharp and clear; alarm, one a quick grating sound like the running down of a broken watch spring, and another a very monotonous, low and plaintive *chee*, hardly ever uttered except when the nest is being visited by an intruder. Song, sweet and plaintive, the latter quality accentuated during the autumn. Local and other names: Redbreast, Robin Redbreast, Robinet, Bob Robin, Ruddock. Sits closely. The male feeds the female very assiduously during the time she is laying and sitting, and during the latter period will sometimes feed the chicks of a different species altogether when he has satisfied the appetite of his mate.

ROBIN.

---

## ROOK.

(*Corvus frugilegus.*)

Order PASSERES ; Family CORVIDÆ (CROWS).

YOUNG ROOK.

*Description of Parent Birds.*—Length from eighteen to twenty-one inches. Beak large, strong, arched towards the point, and black. Round the base of the bill in the adult bird the skin is bare, rough, and light grey. This feature readily distinguishes it from the Carrion Crow. Irides dark brown. The whole of the plumage is black, glossed with rich purple on the upper-parts. Legs, toes, and claws black.

The female is, as a rule, smaller and less brilliant.

*Situation and Locality.*—Amongst the highest branches of tall trees, in colonies or rookeries of various sizes, throughout the country. I have seen colonies of a dozen pairs of birds in an isolated clump of ash trees away up in bleak hilly districts, and as a contrast to this, it may be mentioned that in 1847 it was computed that Newliston Rookery, near Edinburgh, contained no less than 2,663 nests. There is a curious tradition prevalent in different parts of the country to the effect that the birds never establish a rookery on the property of a Dissenter. Rooks are very fickle creatures. They will breed season by season at some well-established haunt, in spite of the fact that their young ones are mercilessly slaughtered every spring, and yet if a few dead trees are removed from a rookery they will take umbrage and desert in a body. Old-established rookeries are sometimes suddenly deserted without any apparent reason whatever. At Thurso, where I secured the picture figuring on this page, Rooks nest quite commonly in chimney pots, although the neighbourhood is by no means destitute of trees.

ROOK SITTING ON HER NEST IN A CHIMNEY POT WITH HER MATE BESIDE HER.

CORNER OF A ROOKERY.

*Materials.*—Sticks and twigs knitted and plastered together with mud and clay, and lined with straw, hay, or wool. The bird is often very particular about the kind of nest it constructs, and will pull it to pieces and rebuild it several times. It is an arrant rogue, and I have watched individuals steal each others' sticks on a good many occasions. The old nests are sometimes repaired in the autumn, and it is said eggs are occasionally laid at that season.

*Eggs.*—Four or five, of a pale green or brownish-green ground-colour, spotted and blotched with greenish- or smoky-brown. *D*istinguished by bird's gregarious habits. Average size about 1.68 by 1.18 in. (*See* Plate I.)

*Time.*—February, March, April, and May; the laying season varying according to the character of the weather.

*Remarks.*—Resident, although numbers increased by arrivals from Continent in winter. Note: *craaw*. Local and other names: White-faced Crow, Craa. Gregarious, and a fairly close sitter.

ROOK.

## RUFF.

(*Machetes pugnax.*)

Order LIMICOLÆ ; Family SCOLOPACIDÆ (SNIPES).

*Description of Parent Birds.*—Length about twelve and a half inches. Bill long, straight, rather slender, and brown. Irides dusky brown. The bird varies very considerably in plumage, one eminent authority having examined two hundred specimens and only found two alike. Yarrell says : "The head, the whole of the ruff or tippet (long plumes growing on the head and neck, and capable of being raised so as to form a kind of shield), and the shoulders of a shining purple-black, transversely barred with chestnut ; scapulars, back, lesser wing-coverts, and some of the tertials, pale chestnut, speckled and tipped with black ; greater wing-coverts nearly uniform ash-brown ; quill-feathers brownish-black, with white shafts ; rump and upper tail-coverts white ; tail feathers ash-brown, varied with chestnut and black ; the feathers of the breast, below the ruff and on the sides, chestnut, tipped with black ; belly, vent, and under tail-coverts white, with an occasional spot of dark brown ; legs and toes pale yellow-brown ; claws black."

The female or Reeve is about two inches less in length, lacks the ruff or tippet altogether, and although not differing much in other plumage, is said to be more uniform in colour as a sex.

*Situation and Locality.*—In a tuft or tussock of some kind of coarse vegetation growing in some wet, swampy place. The bird used to breed at several places in England, but the reclamation of land, the requirements of gourmands, and, later, the greed of the collector, have almost banished it as a breeding species. It has nested during the last year or two in Yorkshire, and occasionally attempts to do so in Norfolk and Lincolnshire. I have seen the Reeve on one or two occasions on the Broads. Our illustrations were secured in Holland by my brother during June, 1905.

*Materials.*—A few bits of dead grass or leaves line the hollow in which the eggs are laid.

RUFFS AT HOME.

*Eggs.*—Four, varying from pale greyish-green to olive-green or olive-brown in ground-colour, blotched and speckled with greyish and rich liver-coloured brown, generally most numerous on the larger end of the egg. Somewhat similar to those of the Redshank, though greyer and not quite so yellow in ground-colour. Size about 1.7 by 1.22 in. (*See* Plate VIII.)

*Time.*—May and June.

*Remarks.*—Migratory, arriving in April and May and departing about September. Notes, *kack, kick.* Local and other names: Reeve (female), Fighting

REEVE'S NEST AND EGGS.

Ruff, Shore Sandpiper, Greenwich Sandpiper, Yellow-legged Sandpiper, Equestrian Sandpiper. Sits closely. The males have a curious habit of meeting at given places, such as that shown in our first illustration, every morning during the pairing season and dancing and fighting until they are selected by the onlooking females as husbands.

---

## SANDPIPER, COMMON.

(*Totanus hypoleucus.*)

Order LIMICOLÆ ; Family SCOLOPACIDÆ (SNIPES).

YOUNG COMMON SANDPIPER.

*Description of Parent Birds.*—Length seven and a half inches. Bill rather long, straight, slender, dark brown towards the tip and lightish brown at the base. Irides dusky brown ; from the base of the beak a light streak runs over the eye and ear-coverts. Crown, back of the neck, back, wing-coverts, and upper tail-coverts greenish-brown, with a line of a darker hue across and down the centre of each feather. Wing-primaries nearly black with dirty white patches on nearly all the inner webs, the secondaries tipped with white. Tail-quills greenish-brown in the centre, barred with greenish-black, the outer webs of the two outside feathers on either side white, barred with greenish-black. Chin, throat, breast, belly, vent, and under tail-coverts white ; sides of the neck and upper portion of the breast duller, and streaked with dark brown

COMMON SANDPIPER ON NEST.

W

or dull black. Legs and toes pale bluish-green; claws dark brown.

The female resembles the male.

*Situation and Locality.*—On the ground; in a hole in a bank, under the shelter of a tuft of grass,

COMMON SANDPIPER'S NEST AND EGGS.

in a tuft of rushes; sometimes in a slight declivity on the bare ground, or in a patch of grass amongst large stones on a little river island; on the banks of rivers, mountain streams with rough, gravelly, and rocky banks, lakes, tarns, and reservoirs, and on the sea shore in the extreme south-west of England (Cornwall, Devon, and Somerset), Wales, the six northern counties of England, Derbyshire, Cheshire,

Shropshire, and Stafford; Scotland, and its surrounding islands, and Ireland. A nest of this species was found a few years ago on the Norfolk Broads.

*Materials.*—Short pieces of dead rushes, sometimes dead leaves, with an inner lining of fine dry grass.

*Eggs.*—Four, pale straw to creamy-yellow in ground-colour, with dark brown spots and blotches on the surface, and underlying markings of light brown and grey. On one occasion I found a nest in North Uist with eggs in it which were marked by dark broad bands of brown round the larger end. Size about 1.5 by 1.08 in. (*See* Plate VIII.)

*Time.*—May and June.

*Remarks.*—Migratory, arriving in April, and departing in September, although individuals may be seen later. Notes: *wheet, wheet, wheet.* Local and other names: Summer Snipe, Sand Lark, Willy Wicket, Sand Lavrock, Spotted Sandpiper. Sitting qualities variable, some individuals sitting closely and others lightly, irrespective of the condition of the eggs.

---

## SANDPIPER, WOOD.

(*Totanus glareola.*)

Order LIMICOLÆ; Family SCOLOPACIDÆ (SNIPES).

The appearance of this bird, even as a visitor, is neither frequent nor regular. It has been found breeding with us in Northumberland and several parts of Scotland, but not during very recent times, so far as I can gather. The eggs number four, of "a pale greenish white ground-colour, speckled and spotted, particularly over the broad end, with dark reddish-brown," according to Yarrell. (*See* Plate VIII.)

## SCOTER, BLACK. *See* SCOTER, COMMON.

---

## SCOTER, COMMON. *Also* BLACK SCOTER.

(*Œdemia nigra.*)

Order ANSERES ; Family ANATIDÆ (DUCKS).

*Description of Parent Birds.*—Length about twenty-one inches. Bill of medium length, swollen into a knob at the base, and flattened at the tip. It is black, with the exception of a ridge of yellow, which commences half an inch from the tip and extends to the base. Irides dusky-brown. The plumage is deep black all over, somewhat glossy about the head and neck. Legs, toes, and webs dusky, darkest on the last.

The female lacks the knob on the bill, and her plumage is not nearly so deep a black in colour.

*Situation and Locality.*—A hollow scraped in the ground or some natural declivity, hidden by low, growing shrubs, sheltering heather, or rushes on small islands, near lochs and rivers, not far from the sea, in the most northern counties of Scotland.

*Materials.*—Bits of dead heather, rushes, and dry grass, with an inner lining of down. The tufts are brownish-grey with pale centres, are large, a little darker than those of the Mallard, and much more so than those of the Goosander.

*Eggs.*—Six to nine, although I have known a case of a bird with thirteen young ones, pale greyish-buff or yellowish-white, sometimes slightly tinged with green, smooth surfaced. Size about 2.5 by 1.78 in.

*Time.*—May and June. I should say from my

COMMON SCOTER'S NEST AND EGGS.

own experience far oftener the latter month than the former.

*Remarks.*—A winter visitor principally. Is said not to breed until it is two years old. Notes: call, a grating *kr-kr-kr*. One authority represents the call-note of the male as *tü-tü-tü-tü*, and the response of the female as *re-re-re-re-re*. Local and other names: Black Duck, Black Scoter, Black Diver. Sits closely, and covers its eggs when leaving them voluntarily.

---

## SHAG.

(*Phalacrocorax graculus.*)

Order STEGANOPODES; Family PELECANIDÆ (PELICANS).

SHAG.

*Description of Parent Birds.*—Length about twenty-seven inches; bill rather long, hooked at the tip, and black. The gape extends behind the line of the eye, and the naked skin about it is black, spotted with chrome yellow. Irides green. The forehead bears a curved-forward kind of crest, which makes its appearance early in the spring. Head and neck, all round, rich dark green, glossed with purple and bronze sheen; back and wing-coverts dark green, with a more intense margin of the same colour round the border of each feather; wings and tail black; breast, belly, and under-parts generally the same as the head and neck; legs, toes, and claws black.

The female resembles the male.

SHAGS' NESTS AND EGGS.

*Situation and Locality.*—Crevices, fissures, and caves in sea cliffs ; sometimes on ledges or amongst the boulders of the rock-strewn beaches of small islands. Pretty generally round our coasts, where suitable accommodation is to be found, but principally on the west coast of Scotland, the Orkneys and Shetlands.

*Materials.*—Seaweed and twigs, lined with grass, the whole plastered and befouled with droppings

YOUNG SHAGS.

and decomposing fish. Where conditions admit, it is a bulky structure.

*Eggs.*—Two to five, generally three ; pale green, almost wholly covered by a chalky substance, which soon becomes discoloured. The eggs resemble those of the Cormorant closely, but are usually a trifle smaller in size. The situation of the nest and presence of birds readily distinguish them. Average measurement, 2.45 by 1.5 in.

*Time.*—May and June.

*Remarks.*—Resident, but subject to local movement. Note: a harsh guttural croak. Local and other names: Crested Shag, Scart, Scarf, Crested Cormorant, Green Cormorant. Shag and Cor-

(See p. 172)

COMMON GULL.
(See p. 162.)

RICHARDSON'S SKUA.
(See p. 364.)

COMMON SKUA.
(See p. 361.)

NOTE.—*In referring to the eggs the above names should be read from left to right*

PLATE 11.

morant are names frequently interchanged by seamen and coast dwellers. Gregarious. A bold and fairly close sitter, sometimes allowing itself to be caught on the nest.

---

## SHEARWATER, MANX.

(*Puffinus anglorum.*)

Order TUBINARES ; Family PROCELLARIIDÆ (PETRELS).

*Description of Parent Birds.*—Length about fourteen inches ; bill rather long, straight, except at the tip, where it is curved downwards, and blackish-brown, lighter at the base. Irides hazel. Head, nape, back, wings, and tail brownish-black ; chin, throat, breast, belly, vent, and under tail-coverts white. The sides of the neck are barred transversely with grey and white. A patch of brownish-black is situated behind the thigh on each side ; legs, toes, and webs flesh-colour, tinged with yellow.

The female is similar to the male, but slightly smaller.

*Situation and Locality.*—At the end of a burrow, generally excavated by the bird, and varying from two to ten or twelve feet in depth, in crevices, and under pieces of rock ; sometimes in a small hole scratched out between two stones. In the Scilly Islands, Wales, on the islands to the west of Scotland, and in suitable places off the Irish coast. It is possible that its peculiar habit of keeping out of sight during the day and only coming forth at night may have conduced to some of its nesting haunts having been overlooked. The bird is known in one case to have been driven away from its nesting stations by Puffins, and in another by rats.

MANX SHEARWATER'S NESTING HOLE.

*Materials.*—Sometimes a few dead fern-fronds, or blades of dried grass, at others nothing whatever.

*Egg.*—One; pure white, smooth, and large for the size of the layer. Size about 2.4 by 1.65 in.

*Time.*—May and June.

*Remarks.*—Resident, but subject to much local movement, except during the breeding season. Notes: a low guttural crooning, delivered in their holes, and when they fly at night time notes that sound like *"It's your fault," "Kitty koo-roo,"* or *"It-y-corka."* Local and other names: Shearwater Petrel, Manx Puffin, Cuckle, Skidden, Scraib, Fachach, Lyrie, Scrapire. Gregarious. A close sitter.

---

## SHELDRAKE. *Also* COMMON SHELDRAKE *or* BURROW DUCK.

(*Tadorna cornuta.*)

Order ANSERES; Family ANATIDÆ (DUCKS).

*Description of Parent Birds.*—Length twenty-four to twenty-six inches; bill fairly long, thick at the base, depressed in the middle, slightly hooked at the tip, and red in colour. On the top of the upper mandible, at the base, is a large fleshy knob. Irides reddish-brown. Head and upper parts of neck dark green; lower half of neck white all round; upper parts of breast and back, rump, wing, and upper tail-coverts white; scapulars and a portion of secondaries blackish; outer webs of inner secondaries rich chestnut. On the last-named feathers is a patch of rich purple-green; primaries almost black; tail-quills white, except at the tips, where they are black; in the middle of the lower breast and belly the feathers are dark brown; sides, flanks, vents, and lower tail-coverts white; legs, toes, and webs flesh-colour.

The female is rather smaller and duller in colour.

*Situation and Locality.*—Rabbit burrows are the favourite haunts of this bird, although it is said sometimes to dig its own burrow or adopt that of a fox or badger; holes under rocks and ruins at various depths, sometimes four or five feet in, at others as many as twelve. In low sand-hills and dunes at various suitable places on the east and west coasts, such as Suffolk, Norfolk, Lincolnshire, Yorkshire, Durham, Northumberland, Lancashire, and Cheshire; in Wales; on various parts of the coast of Scotland, Orkney Islands, Hebrides, and Ireland. Our illustration was obtained in the Hebrides.

*Materials.*—Dry grass, bents, and down from the bird's own body. The tufts are lavender-grey, mixed with a few white ones.

*Eggs.*—Six to sixteen, generally ten to twelve; white, slightly tinged with cream colour. Size about 2.7 by 1.9 in. Nest distinguished by down-tufts.

*Time.*—April, May, and June.

*Remarks.*—Resident. Notes: male call, a deep *korr-korr;* female, a loud quack. Local and other names: Sly Goose, Bargander, Burgander, Burrow Duck, Skeeling Goose, Common Shieldrake, Skel-goose. The nest is very difficult to find; but a good way to accomplish this is to look out for footprints in the sand at the entrance to likely holes, also to watch the movements of the male during flight, and any suspected hole morning and evening when the birds leave and enter. I have generally been astir at three o'clock in the morning for that purpose, and the subject of our picture

SHELDUCK'S BREEDING HOLE.

was discovered not long after that hour by the aid of my binoculars. Sits closely.

---

## SHELDRAKE, COMMON. *See.* SHELDRAKE.

---

## SHOVELLER.

(*Spatula clypeata.*)

Order ANSERES ; Family ANATIDÆ (DUCKS).

*Description of Parent Birds.*—Length about twenty inches. Bill rather long, narrow in the middle, and widening towards the tip, nearly straight, and leaden grey. Irides yellow. Head and upper part of neck deep glossy green, lower part of neck white. Back, in the centre, blackish-brown, the feathers being edged with a lighter tint. Lesser wing-coverts light blue, greater white; scapulars white, quills brownish-black. The speculum, or glossy patch upon the wing, is green. Rump, upper tail-coverts, and tail-feathers brownish-black. Breast and belly chestnut ; vent white ; under tail-coverts black. Legs, toes, and webs reddish-orange; claws black.

The female differs considerably from the male, having her head and neck mottled with two shades of brown ; the feathers on the upper surface of the body are dark brown in the centre, bordered with a lighter shade of the same colour. Under-parts of the body pale brown.

*Situation and Locality.*—In tufts of grass, rushes, and heath beside sluggish rivers, lakes, broads, tarns, and swampy heaths, in favourable situations on the eastern coast of England ; in Wales, Scotland and Ireland.

SHOVELLER'S NEST AND EGGS.

*Materials.*—The little hollow chosen is lined with sedges and dead leaves, dry grass, and, as incubation advances, down. The tufts are of a dark neutral grey colour, lighter in the centre and tipped with white. They and the female form reliable distinctions.

*Eggs.*—Seven to ten, or occasionally fourteen, buffish-white, tinged with green, unspotted, and slightly polished. Closely resemble those of the Mallard and Pintail. Size about 2.0 by 1.5 in.

*Time.*—May.

*Remarks.*—Resident and migratory, being more numerous in winter than in summer. Notes: *quack*, uttered in deeper tones by the male; when flying, *puck, puck*. Local and other names: Broad Bill, Blue-winged Shoveller. Sits closely.

---

## SHRIKE, RED-BACKED.

(*Lanius collurio.*)

Order PASSERES; Family LANIIDÆ (SHRIKES).

YOUNG RED-BACKED SHRIKE.

*Description of Parent Birds.*—Length about seven and a half inches; bill rather short, hooked at the tip, and black or dusky brown, lighter at the base of the lower mandible. Near the end of the upper mandible is a prominent tooth or notch. Irides hazel. Three or four strong black bristles spring from just above the gape. Round the base of the upper mandible, through the eyes,

RED-BACKED SHRIKE'S NEST AND EGGS

and as far as the ear-coverts, the feathers are black; crown and back of neck grey; back and wing-coverts bright reddish-brown; wing-quills dull black margined with reddish-brown; upper tail-coverts reddish-grey; tail-quills, in centre, black tipped with white, rest white on the basal half, and black from thence to the end, which is slightly tipped with white; chin grey; breast, belly, and under-parts of a rosy tinge, with the exception of the under tail-coverts, which are white; legs, toes, and claws dusky-black.

The female is very much less conspicuous in her plumage. Her beak is not so dark in colour; over her eye is a yellowish-white streak; her upper parts dull rusty-brown, tinged with grey on the neck and tail-coverts; chin, throat, breast, and under-parts greyish-white, barred with greyish-brown.

*Situation and Locality.*—In high rough hedges, thorn bushes in woods, and on rough commons. Our illustration is from a photograph of a nest in a slight thorn bush, surrounded by hazels and big trees in a small Surrey spinney, where I meet with a nest every year regularly. I found a Red-backed Shrike's nest in a low bramble bush, intermixed with rushes, a few years ago close to London. The nest was not more than eighteen inches from the ground and within a few feet of a much-used country lane. The cock bird was so bold that he came within four feet of me as I stood looking at his mate sitting upon her eggs. Breeds pretty generally over England, with the exception of the extreme north, and in Wales, but is rarely met with in Scotland or Ireland.

*Materials.*—Slender twigs, dead grass, stalks, dead weeds, honeysuckle stems and stalks, roots, wool, moss, and occasionally feathers, lined with hair, sometimes with willow catkins and fine, fibrous roots. As a rule it is a very large nest for the size of the bird; but I have noticed that specimens differ in this respect as well as in the character of the materials employed in their construction.

*Eggs.*—Four to six, generally four or five; very variable in ground-colour and markings; pale buffish-white, spotted, freckled, and blotched with pale reddish-brown, and underlying markings of grey or salmon-colour, marked with light red and lilac-grey. Some varieties are white, greyish-white, yellowish-white, or greenish in ground-colour. As a rule, the markings form a ring round the larger end. Size about .9 by .66 in. (*See* Plate III.)

*Time.*—May and June. I once found one at the beginning of July.

*Remarks.*—Migratory, arriving in May and departing in August or September. Note: call, a harsh croak; song a mixture of the notes of the Goldfinch, Blackcap, Nightingale, and other birds frequenting its vicinity according to Bechstein. Local and other names: Jack Baker, Murdering Pie, Whiskey John, Butcher Bird, Flusher, Cheeter. A fairly close sitter.

RED-BACKED SHRIKE ON NEST

## SHRIKE, WOODCHAT.

(*Lanius pomeranus.*)

Order PASSERES ; Family LANIIDÆ (SHRIKES).

A rare and accidental visitor, which is said to have bred once or twice in the Isle of Wight.

The eggs number four or five, and are, according to Yarrell, very variable in colour, "some being white tinged with green, or pale olive blotched irregularly with olive and lilac of different shades, the markings sometimes diffused and sometimes forming regular spots often disposed in a zone, while other specimens are of a cream colour with light red and suffused lilac spots. Size, .97 to 1.86 by from .7 to .65 in. (*See* Plate III.)

---

## SISKIN.

(*Chrysomitris spinus.*)

Order PASSERES ; Family FRINGILLIDÆ (FINCHES).

*Description of Parent Birds.*—Length nearly five inches. Bill short, conical, sharp-tipped, and orange-brown. Irides dusky-brown. Top of head black ; over and under each eye is a yellowish streak. Sides of head yellowish-green ; nape and wings (except greater coverts and quills, which are brownish-black, tipped and bordered with yellow) greenish-olive, streaked with black ; rump yellow ; upper tail-coverts greenish-olive. Tail slightly forked, and dusky-black, yellowish on the upper half, with the exception of the middle pair of feathers. Chin black, throat and breast yellowish-green ; belly, sides, flanks, vent, and under tail-coverts, greyish-white,

SISKIN'S NEST AND EGGS

streaked with dusky-black. Legs, toes, and claws brown.

The female is smaller, and lacks the black on the crown and chin. Her upper-parts are olive-brown, throat and breast greenish-yellow, and rest of under-parts greyish-white. With the exception of the centre of the belly she is streaked all over with dusky-black.

*Situation and Locality.*—In plantations, woods, and forests. Its nest has been found, on very rare occasions in different parts of England, in furze and juniper bushes; but in Scotland, where it breeds sparingly, it adopts higher situations amongst the forks and branches of fir-trees. It breeds regularly in the south of Ireland, where, thanks to our friend Mr. Ussher, the accompanying photograph was obtained.

*Materials.*—Slender twigs, dried grass, and moss, lined internally with hair, rabbit or vegetable down, and sometimes a few feathers.

*Eggs.*—Four to six, greyish-white, tinged with green or pale bluish-green, spotted and speckled with rusty and dark brown spots, sometimes streaked with the darker colour. The markings are generally scattered over the surface of the eggs, but are sometimes collected round the larger ends. They resemble those of the Goldfinch very closely indeed, but are said to run larger, and the ground-colour to be of a darker tinge. The situation of the nest and a sight of the owner are the only reliable evidences, however. Size about .66 by .52 in. (*See* Plate III.)

*Time.*—April, May, and June.

*Remarks.*—A winter visitor of erratic appearance; a few resident. Notes: call, a metallic *keet;* alarm note, *chuck-a-chuck, keet.* Some naturalists represent the call note as a weak *tit-tit-tit-tit*, and *tsyzing*, others as a loud *deedel* or *deedlee*. Local and other names: Aberdevine (used by bird-catchers), Barley Bird. Nest difficult to find. A close sitter.

---

## SKUA, COMMON. *Also* SKUA, GREAT.

(*Stercorarius catarrhactes.*)

Order GAVIÆ; Family LARIDÆ (GULLS).

COMMON SKUA GETTING READY TO ATTACK.

*Description of Parent Birds.*—Length about twenty-four inches. Bill of medium length, hooked at the tip, and with bare skin round its base, black. Irides dark brown. Head and neck dark umber-brown, slightly streaked with lighter brown; back, wings, and tail-coverts dark brown, streaked with light reddish-brown. In some specimens the feathers at the nape, and the middle and edges of those on the back, are greyish-white. The wing-quills are white at the base and blackish-brown towards the tip; tail-quills very dark brown. Chin and front of neck, breast, belly, vent, and under tail-coverts dusky rust-colour. Legs, toes, and webs black; claws large, much curved, strong, and black.

*Situation and Locality.*—On the ground, amongst moss, coarse short grass, and stunted heather; at two places in the Shetland Islands only, where the birds and their nests and eggs are protected, and

NEWLY HATCHED COMMON SKUA.

COMMON SKUA'S NEST AND EGGS.

it is gratifying to be able to say they have, during the last decade or two, increased in numbers—on Herma Ness in Unst, at any rate.

*Materials.*—All the nests I examined in the Shetlands some years ago were made of dead grass and bents in varying quantities. In nearly every case a mock nest had been built not far from the one containing eggs, and the birds have been known, when an original nest became flooded through heavy rains, to convey their eggs to the supplementary structure.

*Eggs.*—Two, occasionally only one, varying from light buff to dark olive-brown, blotched and spotted with dark brown and rusty- or greyish-brown. Similar to Lesser Black-backed and Herring Gulls, but markings are fewer and duller, and presence of parent birds readily settles the point. Size about 2.85 by 1.95 in. (*See* Plate XI.)

*Time.*—May and June.

*Remarks.*—Migratory, arriving at its breeding haunts in April and leaving in August. Notes: *ág-ag* and *skua*. Local and other names: Great Skua, Bonxie, Brown Gull, Skua Gull, Morrel Hen. Gregarious. Sits lightly and readily attacks a single human intruder approaching its nest and eggs or young. The method of attack is to swoop from a considerable height at a terrific speed, and, when sufficiently close to the intruder, to drop both feet and hit him with their fronts on the back of the head. As soon as the blow has been delivered the birds shoot upwards, and circle round for a fresh attack. According to my experience they never strike the object of their resentment in the face.

## SKUA, GREAT. *See* SKUA, COMMON.

---

## SKUA, RICHARDSON'S.

(*Stercorarius crepidatus.*)

Order GAVIÆ; Family LARIDÆ (GULLS).

RICHARDSON'S SKUA SITTING.

*Description of Parent Birds.*—Length about twenty inches. Bill moderately long, strong, straight, except at the tip, where it is hooked, bluish lead-colour at the base, and blackish elsewhere. Irides dark brown. This bird is subject to considerable individual variation, and there are two distinct and well-marked varieties, known as "light" and "dark" which interbreed freely. The dark variety is more common in low latitudes, and the light one in high latitudes, as might be expected.

Mr. Seebohm, in describing the bird, says: "In the adult of the dark form, the whole of the plumage is an almost uniform dark sooty-brown, slightly suffused with slate-grey on the upper-parts, and with a bronzy yellow on the sides of the neck.

"In the adult of the light form, the slate-grey of the upper-parts is a little more pronounced than in the dark form. The general colour of the under-parts is white, shaded with brown on the sides of the breast, the vent, and the under tail-coverts; the white on the throat extends round the sides

NEST AND EGGS OF RICHARDSON'S SKUA

of the neck and across the lower ear-coverts, almost to the nape, and is suffused with yellow. Legs and feet black."

The female is, so far as is known, indistinguishable from the male, except that the elongated feathers of the tail are somewhat shorter.

*Situation and Locality.*—On the ground, amongst short heather, moss, and coarse grass, in the moorland parts of the Orkneys, Shetlands, Hebrides, and in one or two places on the extreme northern mainland of Scotland. Our illustrations are from photographs taken in the Outer Hebrides.

*Materials.*—Dried grass and moss, used as a scanty lining to the hollow in which the eggs are laid; sometimes none at all.

*Eggs.*—Two; as many as three have been found, and upon occasions only one. Varying in ground-colour from dark olive-green to brownish-green, irregularly spotted and blotched with differing shades of dark brown and greyish-brown, generally distributed over the entire surface of the egg, but sometimes most numerous at the larger end. They closely resemble some of the Gulls in appearance, and the only safe method of identification is to watch the parent bird on or off the nest, a by no means difficult task with a pair of ordinary field-glasses. Size about 2.3 by 1.62 in. (*See* Plate XI.)

*Time.*—May and June.

*Remarks.*—Migratory, arriving in May and departing in August and September. Notes: *mee* and *mee-awk*, represented by some authorities as *kyow* and *yah-yah*. Local and other names: Arctic Gull, Arctic Skua (also applied to Long-tailed or Buffon's Skua), Shooi, Scoutie-Allen, Black-toed Gull. Sits lightly.

## SKYLARK.

(*Alauda arvensis.*)

Order PASSERES ; Family ALAUDIDÆ (LARKS).

YOUNG SKYLARK.

*Description of Parent Birds.* — Length about seven inches. Bill of medium length, straight, strong, and dark brown. Irides hazel ; crown dark brown, the feathers being edged with a lighter and redder tinge, and somewhat elongated, forming a crest which is erectable at will. Back of neck, back, wing, and tail-coverts reddish-brown, each feather being bordered with a pale tint. Wing and tail-quills dusky-brown, with lighter edges and tips. Throat and breast light cream-colour, spotted with dark brown; under-parts pale straw-colour, tinged with brown on the thighs and flanks. Legs, toes, and claws brown; middle toe largest, and hind claw very long and curved.

The female is not quite as large as the male, but is similar in her plumage.

*Situation and Locality.*—Under tufts of grass, ling, and heath, sometimes on the plain open ground, in a slight declivity. In cultivated and uncultivated districts throughout the United Kingdom, but not in woods and plantations.

*Materials.* — Grass, roots, and horsehair, the latter two often quite absent and the first used sparingly.

SKYLARK'S NEST ON THE CROWN OF A FURROW.

SKYLARK'S NEST IN A ROUGH GRASS PASTURE

*Eggs.*—Four to five, of a dirty white ground-colour, occasionally tinged with olive-green, thickly spotted and speckled with olive-brown, and underlying markings of greyish-brown. The markings are generally so thickly and evenly distributed as to hide the ground-colour, but occasionally the markings are less thickly distributed and collected in a kind of belt at the larger end of the egg. Distinguished from Woodlark by crowded olive-brown markings. Size about .93 by .68 in. (*See* Plate II.)

*Time.*—April, May, June, and July.

*Remarks.*—Resident, though subject to partial migration and much local movement. Notes: song consists of several strains, trilling, warbling notes, variously modulated, and interrupted now and again by loud whistling. At the beginning of the breeding season the song is of short duration, and towards the end is frequently uttered whilst the bird is on the ground. Local and other names: Lavrock, Field Lark. A close sitter when the ground is rough and uneven, but not particularly so when it is bare and the situation exposed.

SKYLARK FEEDING YOUNG.

## SNIPE, COMMON.

(*Gallinago cœlestis.*)

Order Limicolæ ; Family Scolopacidæ (SNIPES).

YOUNG COMMON SNIPE.

*Description of Parent Birds.*—Length about ten and a half inches ; beak very long (about two and three-quarter inches), straight, and pale reddish-brown at the base and dusky towards the tip. Irides dark brown. Crown blackish-brown, divided in the centre by a buffish-brown longitudinal line ; another line of the same colour commences at the base of the bill and passes over the eyes. A dusky line passes from the base of the beak to the eye ; back and scapulars dark brown, barred with rusty-brown. Four distinct lines of dark brown feathers, bordered with rich buff, run along the upper-parts of the body. Wing-coverts dull black, spotted with pale brown and tipped with white ; quills dull black, some of them edged and others tipped with white ; upper tail-coverts dusky-black, barred with brown ; tail-quills black, barred and spotted with dull orange-red, and tipped with pale reddish-yellow ; chin brownish-grey ; neck and cheeks light brown ; front and sides of neck a mixture of dark and rusty-brown ; breast, belly, and vent white ; under tail-coverts pale brown, barred with dusky-black ; legs and toes greenish-brown, dusky, or leaden colour.

COMMON SNIPE'S NEST AND EGGS.

COMMON SNIPE: MALE AND FEMALE COVERING CHICKS.

The female is practically like the male, except that she is a trifle larger.

*Situation and Locality.*—On the ground, in a tuft of long coarse grass, amongst rushes or heather, generally hidden by an overhanging tuft of half-dead grass. In wet pasture-lands, marshes, and swamps, near tarns and bogs, in suitable localities throughout the United Kingdom.

*Materials.*—A few dry grass stalks, slender sprigs of dead heather, or other bits of herbage, used as a lining; sometimes hardly anything at all.

*Eggs.*—Four; ground-colour varying from olive-green to greyish-yellow; spotted and blotched with blackish-brown, light brown, and underlying markings of grey. The markings are generally most numerous at the larger end, and the eggs are sharply pointed at the smaller. Size about 1.58 by 1.1 in. (*See* Plate IX.)

*Time.*—April and May, although nests containing eggs have been found as early as the third week in March, and I have found them quite fresh as late as the end of July.

*Remarks.*—Resident, but subject to local migration. Notes: *tjick-tjuck, tjick-tjuck,* uttered both whilst the bird is perched and between the drummings when it is on the wing at dusk and on dull days. Local and other names: Hammer Blate, Whole Snipe, Heather Bleater. Sits closely, and simulates lameness when flushed, in order to draw the intruder away from its eggs. The male bird helps the female to feed and take care of the chicks, as I proved when I photographed both birds covering their offspring, figured on the previous page.

(*See p.* 409.)

COMMON GUILLEMOT.
(*See p.* 158.)

COMMON TERN.
(*See p.* 405.)

LITTLE TERN.
(*See p.* 408.)

NOTE.—*In referring to the eggs the above na nes shou*

PLATE 12.

## SPARROW, COMMON. *Also* SPARROW, HOUSE.

(*Passer domesticus.*)

Order PASSERES ; Family FRINGILLIDÆ (FINCHES).

MALE COMMON SPARROW.

*Description of Parent Birds.*—Length nearly six inches. Bill short, thick at the base, strong, and dusky. Irides hazel. Crown and nape ash-grey (where Tree Sparrow is chestnut-brown) ; back, scapulars, and wing-coverts reddish-brown, mixed with black, the last-named feathers being tipped with white, which forms a bar across the wing. Wing-quills dusky, edged with reddish-brown ; tail dusky-brown, bordered with grey. Cheeks whitish ; chin and throat black ; belly and vent light ash-grey. Legs and toes brown ; claws black.

The female is not quite so large as the male ; the plumage on her upper-parts is not so bright, and she lacks entirely the black on the chin and throat. Town-dwelling birds have their plumage much dulled by grime and smoke, which even penetrates to the interior of their bones.

*Situation and Locality.*—Holes in walls, under tiles and slates, behind signboards fastened against walls, holes in cliffs, the old nests of Sand and House Martins ; in the thatch of houses, barns, and ricks, holes in hollow trees ; amongst ivy trained against houses ; the rafters of stables and sheds ; amongst the loose sticks under Rooks' nests, in old Magpie's nests, Pigeon cotes, on ledges, amongst the branches of trees, and almost any situation capable of accommodating a few straws and feathers. I know several small villas in the north of London with globular cast-iron ornaments on their summits; into these the Sparrows have found their way and turned them to account as nesting sites. I met once with a colony in a low whitethorn hedge, quite away from any houses whatever, and have counted as many as twenty-six nests in the branches of a single tree. Our full-page illustration represents one of two Pigeon cotes, standing close together, out of which upwards of two hundred Sparrows' eggs were taken during a single nesting season.

*Materials.*—Straw, hay, bits of string, moss, worsted, and cotton rags, wool and hair, with a liberal inner lining of feathers. Where the nest is under cover it is not so bulky, and is open at the top, as a rule; but where it is exposed it is covered, bulkier, and better constructed, with a hole in the side, and generally near the top.

*Eggs.*—Four to seven, generally five or six, pale grey or greyish-white, sometimes tinged slightly with green or blue, spotted and blotched thickly with brown, of various shades, and grey. I have seen Sparrows' eggs pure white. One egg generally differs from the others in a clutch in regard to the character of its markings. Distinguished from Tree Sparrows' eggs by larger size, situation, and female lacking black patch on her chin. Size about .9 by .6 in. (*See* Plate II.)

*Time.*—March to July or August.

*Remarks.*—Resident. Notes: a monotonous chirrup, and a hurried scolding when engaged in

DOVECOTE IN WHICH COMMON SPARROWS NESTED.

warfare. One authority makes the surprising statement that when under proper training the bird can be taught to sing even better than the Canary. Local or other name: none. Sits pretty closely.

YOUNG COMMON SPARROW.

---

## SPARROW, HEDGE. *Also* HEDGE ACCENTOR.

(*Accentor modularis.*)

Order PASSERES; Family SYLVIIDÆ (WARBLERS).

HEDGE SPARROW ON NEST.

*Description of Parent Birds.*—Length about five and a half inches; bill of medium length, almost straight, light brown at the base, and darker at the tip. Irides reddish-brown. Crown and nape dull bluish-grey, streaked with brown; back and wings dusky-brown, the feathers being edged with reddish-brown; wing-quills dusky-brown; tail-coverts olive-brown; tail-quills dusky-brown and slightly forked; chin, throat, sides of neck, and upper-parts of breast

HEDGE SPARROW'S NEST AND EGGS.

dark bluish-grey; breast and belly buffish-white; sides pale yellowish-brown, streaked with a darker tinge of the same colour; vent and under tail-coverts pale tawny-brown; legs and toes dark orange-brown; claws black.

The female is duller in plumage, with more markings on the head and sides, and is a trifle smaller.

*Situation and Locality.*—Hawthorn hedges are favourite situations; the nest may, however, be found in all kinds of low bushes, such as furze, gooseberry, briars, brambles, and nettles. On one occasion I met with three nests, containing eggs, close together amongst the black-currant bushes of a small garden in Westmorland; and remember on another finding one quite on the ground in Yorkshire. Found pretty generally throughout the United Kingdom, with the exception, perhaps, of the islands lying to the north, and some of the smallest and bleakest of those to the west of Scotland.

*Materials.*—Slender twigs (sparingly used and sometimes entirely absent), roots, moss, and dry grass, with an inner lining of wool, hair, and feathers. I have on several occasions seen nests made entirely of moss and cowhair. Many individuals of this species have a habit of covering their eggs when they leave them before commencing to sit. In July, 1905, I found a Cuckoo's egg covered over in a Hedge Sparrow's nest which contained no other eggs.

*Eggs.*—Four to six, of a beautiful unmarked turquoise-blue. Size about .77 by .6 in. (*See* Plate V.)

*Time.*—March, April, May, and June, sometimes as late even as July.

*Remarks.*—Resident. Notes: a low plaintive *cheep-cheep*, and a cheerful, though not long sustained, song. Local and other names: Hedge Accentor, Shufflewing, Hedge Warbler, Dunnock, Hempie, Cuddy. Sits closely, and slips away without demonstration.

HEDGE SPARROW FEEDING YOUNG.

## SPARROW, HOUSE. *See* SPARROW, COMMON.

## SPARROW, REED. *See* BUNTING, REED.

## SPARROW, TREE.

(*Passer montanus.*)

Order PASSERES; Family FRINGILLIDÆ (FINCHES).

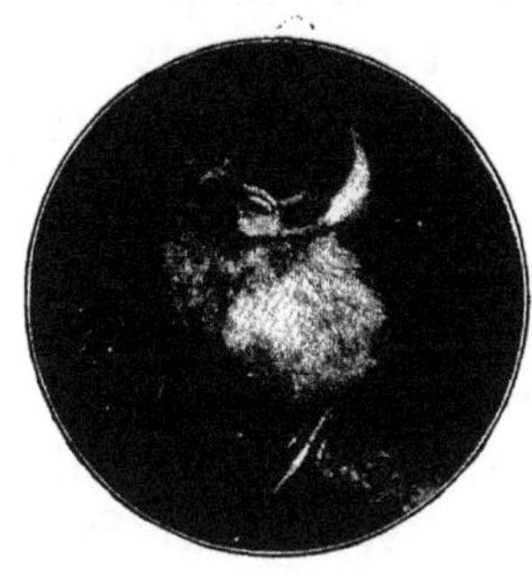

YOUNG TREE SPARROW.

*Description of Parent Birds.*—Length about five and a half inches. Bill short, strong, broad at the base, and lead coloured. Irides hazel. Crown and nape dull chestnut-brown (where Common Sparrow is ash-grey). Beneath the eye is a streak of black. Cheeks white, with a large black spot. The upper part of the back and scapulars bright rusty-brown, streaked with black; lower part of back and upper tail-coverts brownish-grey. Lesser wing-coverts bright rusty-brown; greater black, with rusty-coloured edges and white tips; quills dull black, bordered with rusty-brown. Tail-quills greyish-brown, edged with lemonish-brown. Chin and throat black; sides of neck, running somewhat far back, white; breast bright ash-grey; belly dull white, tinged with buffish-brown on the sides, vent, and under tail-coverts. Legs, toes, and claws pale yellowish-brown.

The female is a little smaller, but her plumage differs in nothing but its lesser brilliancy from that of the male. In this respect the species differs radically from the House Sparrow, the female of which is not adorned by the black patch on the chin and throat, so conspicuous in the male. In the case of the Common Sparrow the young are all like the female in their first coat of feathers, but the chicks of the species under notice have the chin and throat black as in the case of their parents.

*Situation and Locality.*—Holes in pollards and other trees, crevices of rocks, holes in walls, and the thatch of barns. It is very local, and in few places numerous. It is said to breed most commonly in the midland and eastern counties of England, but I have met with it far more numerously in the Outer Hebrides than anywhere else. As an example of its absolute commonness in some parts of the Western Isles I may mention that I one day found eight nests, five containing eggs and three young ones, all in the ruins of one small building. Whilst at St. Kilda in 1896 I succeeded in finding four Tree Sparrows' nests. It is met with sparingly on the mainland of Scotland, in Wales, and Ireland.

TREE SPARROW'S NESTING SITE.

*Materials.*—Straws, dry grass, and roots, lined with hairs and liberal supplies of feathers. It is generally open at the top, but domed when situation demands it.

*Eggs.*—Four to six, although I once found eight in a nest, and a friend of mine at Northampton has seen a clutch of nine. Generally four or five, greyish-white in ground-colour, thickly spotted all over with dark grey or dark brown. Occasionally the ground-colour is white, thickly spotted and freckled with grey spots and blotches. One member of a clutch is often lighter coloured than the rest, and sometimes the eggs are streaked with a dark line or two. They are not unlike the eggs of the Pied Wagtail or Meadow Pipit, but of course the position of nest differs widely. Size about .8 by .57 in. (*See* Plate II.)

*Time.*—April, May, June, and July.

*Remarks.*—Resident, but subject to local movement. Notes: numerous Sparrow-like chirrups. Local or other name: Mountain Sparrow. A fairly close sitter.

TREE SPARROW WITH FOOD FOR YOUNG.

## SPARROW-HAWK.

(*Accipiter nisus.*)

Order ACCIPITRES ; Family FALCONIDÆ (FALCONS).

SPARROW-HAWK AT NEST.

*Description of Parent Birds.* — Length about twelve inches ; beak short, curved, and bluish ; bare skin round the base of the beak yellow. Irides yellow. Head, nape, back, wing, and upper tail-coverts deep bluish-grey, edged with rusty-red ; wing-quills dusky, barred with black on the outside webs, and spotted with white on the lower portions of the inside webs ; tail deep ash-colour, crossed with broad bars of dull black and tipped with whitish-grey ; throat, breast, sides, belly, and vent reddish-brown, marked with transverse bars of orange in some and brown in others ; legs and toes yellow ; claws black.

The female is about three inches longer, and nearly twice as heavy. Her upper-parts are browner, with the exception of the back of the head, which is greyer. The breast and under-parts are lighter, and the markings on them larger and browner. Both sexes are subject to considerable variation, and are said to grow greyer with age.

*Situation and Locality.*—In fir, alder, larch, oak, pine, and other trees, in well-wooded districts throughout the British Isles. It is generally placed in a fork or on a strong horizontal branch.

*Materials.*—Sticks and twigs, the finest in the

SPARROW-HAWK'S NEST AND EGGS.

centre, which is simply a slight hollow on a large platform. Many naturalists assert that the bird often utilises the old nest of a Magpie or Crow; but Mr. Dixon says that the nest is always made by the birds themselves. I have examined a great number of nests in widely different parts of the country, and in nearly every instance I am able to endorse what he says. Besides, I have actually photographed a bird of this species in the act of adding sticks to her nest, which is proof of her capacity for nest building.

*Eggs.*—Four to six, generally five; ground-colour white, tinged with blue or bluish-green, clouded, blotched, and spotted with pale brown and dark rich brown. The markings generally form a zone round the larger end of the egg; sometimes the ground-colour is almost entirely hidden, and at others nearly, if not quite, all exposed. Size about 1.65 by 1.3 in. (*See* Plate VI.)

*Time.*—April, May, and June.

*Remarks.* — Resident. Note: alarm, a harsh scream. Local or other name: Pigeon Hawk. A close sitter.

YOUNG SPARROW-HAWKS IN NEST

## STARLING.

(*Sturnus vulgaris.*)

Order Passeres ; Family Sturnidæ (STARLINGS).

STARLING.

*Description of Parent Birds.*—Length about eight and a half inches. Bill rather long, nearly straight, and yellow, except at the base, where it is light bluish-grey. Irides brown. The head, neck, and upper parts are black, glossed with purple-green and steel-blue; the feathers of the head and neck are very slightly tipped with buffish-white; those of the back, rump, and upper tail-coverts are tipped with larger spots of the same colour. Wing and tail-quills greyish-black, edged outwardly with buffish-white; breast and belly black, glossed with purple and steel-blue; vent and under tail-coverts black, tipped and edged with buffish-white, lighter than on the back. Legs, toes, and claws reddish-brown.

The female is not so bright as the male, either in her plumage or bill.

*Situation and Locality.*—Fissures and crevices in cliffs, holes in the gables of old houses, stables, and barns; in ruins, under eaves, in hollow trees, holes in the ground on steep hill sides, and sometimes even amongst the loose sticks forming the foundation of a Rook's nest or Osprey's eyrie. Our full-page illustration represents a hollow apple-tree, in

STARLING'S NESTING HOLE AND NEST AND EGGS.

which a pair of Starlings breed every year. During one season three clutches of six, six, and five eggs respectively were taken out of this nest, on account of the owner of the orchard having seen the parent birds feeding from the fruit of a strawberry-bed close by. I have known Starlings on several occasions adapt the nests of House Sparrows when they have been built in thick Scotch fir trees, and once whilst staying at an hotel in Kinross I watched a pair of birds of this species feeding a brood of young ones in a nest which was situated in a thick evergreen where one might have expected the home of a Song Thrush or a Blackbird. Indeed I am not sure that the chicks were not occupying an old nest built by one of the above-mentioned birds.

*Materials.*—Straw, hay, and fibrous roots, lined with feathers, wool, moss, or whatever may be easily obtainable. Nests made with nothing whatever but straw are often met with.

*Eggs.*—Four to six, of a uniform pale blue. I have seen clutches once or twice that were as near white as possible. Size about 1.18 by .84 in. (*See* Plate I.)

*Time.*—April, May, and June, although eggs have been seen in January and later than June.

*Remarks.*—Resident, but subject to southern movement in winter. Notes: alarm, *spate, spate;* song, a mixture of all kinds of sounds, the bird being a very clever imitator. Local and other names: Sheep Starling, Stare, Sheep Stare, Brown Starling, Starnel, Sheep's Starnel. A close sitter.

We have so many birds of this species in our country nowadays, in consequence of a long succession of mild, open winters, that I am of opinion that some of them do not succeed in finding nesting quarters. On May 1st, 1905, I was staying with some friends near the Blackwater in Essex, and made the following entry in my diary: "In the evening I visited a little wood where Starlings roost in some thick bushes growing alongside tall trees, and although late for flocks of unpaired birds, found thousands of Starlings at roost. When I struck the bushes with my walking stick, the birds rose in a black cloud, and the noise made by their hurrying wings was just like that produced by the shooting of truckloads of small coals into the hold of a steamer. The air was quite still, and I could plainly smell the aroma thrown off the bodies of the winged throng."

---

## STONECHAT.

(*Pratincola rubicola.*)

Order PASSERES; Family TURDIDÆ (THRUSHES).

MALE STONECHAT.

*Description of Parent Birds.*—Length a little over five inches. Bill of medium length, slightly curved downward, and black. Irides dusky-brown. Head, nape, and back black, edged with tawny-brown; rump and upper tail-coverts white, tipped with tawny-brown and black. Wing-coverts black, edged and tipped with rusty-brown; those nearest the body are white, and form a conspicuous patch on the wings; quills dusky, some of them edged with rusty-brown. Tail-feathers black, faintly edged and tipped with pale reddish-brown. Chin and throat black; sides of neck white; breast dark rich rust colour, belly much lighter; vent and under tail-coverts a mixture of black and white, which varies in individual specimens; some are dark and others quite light coloured in these parts.

The female is dull brown on the head, nape, and back, the feathers being edged with buff; the rump is brownish, the chin buff, the sides of the neck brownish-white, and the breast and belly duller.

*Situation and Locality.*—On or near the ground, amongst grass, brambles, at the foot of gorse bushes, and amongst rough, tangled vegetation; in pastures, grass fields, on furze and heath-covered commons, and ground covered with juniper brambles, boulders, and bushes. The nest is somewhat difficult to find; the one shown in our illustration was stumbled upon quite by accident on a Suffolk common. The bird is very local, but breeds more or less in suitable localities all over the British Isles.

*Materials.*—Roots, moss, and dry grass, with an inner lining of hair, feathers, finer grass, and sometimes a little wool.

*Eggs.*—Four to six, rarely seven, of a pale bluish-green, closely mottled, and especially round the larger end, with reddish-brown spots. Sometimes without any spots at all. Distinguished from the eggs of the Whinchat by lighter ground-colour and more defined markings, also by parent birds. Size about .7 by .57 in. (*See* Plate IV.)

*Time.*—April and May.

*Remarks.*—Resident, but subject to local migration. Notes: *ŭ-tic, ŭ-tic,* changing when the young are hatched to *chuck, chuck.* Local and other

STONECHAT'S NEST AND EGGS.

names: Stoneclink, Stone Chatter, Stone Smick, Stone Chack, Stonesmith, Chick Stone, Black Cap, Moor Titling (a name generally applied to the Meadow Pipit in some districts). A fairly close sitter, but when at the foot of a furze bush the bird runs for some distance before taking flight. It is extremely wary, and I have lain for hours and hours together watching a pair through my binoculars without being able to discover the nest.

---

## SWALLOW.

(*Hirundo rustica.*)

Order PASSERES; Family HIRUNDINIDÆ (SWALLOWS).

SWALLOW.

*Description of Parent Birds.*—Length about eight and a half inches; bill short, straight, somewhat flat, and black. Irides hazel. Forehead chestnut; crown, neck, back, rump, and upper tail-coverts steely-blue; wings and tail sooty-black, the former long and sword-like, and the latter very much forked. From several specimens examined I have found that one side of the tail (sometimes the right and at others the left) is a trifle longer than the other; chin and throat chestnut. Round the lower part of the throat and upper part of the breast is a broad steely-blue collar; lower breast, belly, vent, and under tail-coverts buffy-white; legs, toes, and claws short, slender, and black.

The female is not so richly marked in her plumage, and her tail is not so long.

*Situation and Locality.*—Generally, as shown in

SWALLOW'S NESTING-SITE AND NEST AND EGGS.

our full-page illustration, on the rafter of a barn, stable, or shed. Sometimes on ledges and other projections in chimneys and from walls. I recollect once finding one inside an old disused mountain lime-kiln, and have known odd pairs of birds to breed underneath the arches of stone bridges. Not long ago I saw one, built upon a piece of ivy dangling from the roof of a shed through which it had grown.

*Materials.*—Mud, straws, dry grass, and feathers in liberal quantities.

*Eggs.*—Four to six, generally five, white, speckled, and blotched with dark red-brown, and underlying markings of ash-grey. The markings are generally most numerous round the larger end. I have seen eggs once or twice with hardly any marking on at all. Size about .83 by .55 in. (*See* Plate V.)

*Time.*—May, June, July; and sometimes eggs may be met with as late as the beginning of August.

*Remarks.*—Migratory, arriving in April and departing in September and October. Notes: *wet-wet*, a warbling kind of song note, and *pink*, *pink* when the bird is alarmed. Local and other names: Barn Swallow, House Swallow, Chimney Swallow, Common Swallow. Not a very close sitter until incubation has advanced some stages.

YOUNG SWALLOWS.

## SWAN, MUTE.

(*Cygnus olor.*)

Order ANSERES; Family ANATIDÆ (DUCKS).

*Description of Parent Birds.*—Length about four feet eight inches to five feet; beak fairly long, black on the edges and tip, rest red. The

SWANS AT NEST.

knob, or tubercle, on the base of the upper mandible, and the naked skin between the eyes and the base of the bill, black. Irides brown. The whole of the plumage is snowy white. Legs, toes, and webs black.

The female has the knob smaller, the neck more slender, and swims deeper in the water.

*Situation and Locality.*—On the ground amongst reeds and coarse vegetation. On small islands and banks. Close to the water of sluggish rivers or lakes in various parts of the country, but principally on the Thames, Avon, Norfolk Broads, and at Abbotsbury in Dorsetshire.

*Materials.*—Reeds, rushes, dry flags, and grass, often in great quantities, and down.

*Eggs.*—Three to twelve, generally six or seven; dull greenish-white. Size about 4.5 by 2.9 in.

*Time.*—March, April, and May.

*Remarks.*—Strictly speaking, this bird has no proper claim to be included in a work of this character, for although it breeds in a perfectly wild state on the Continent, it has never been known to do so within the limits of the British Isles. The case of the Pheasant, however, another introduced half-domesticated bird, holding its own only through strict protection, paves the way.

The Mute Swan is said to have been first intro-

SWAN'S NEST AND EGGS.

duced into this country from Cyprus by Richard I., who commenced to reign in 1189. It is considered a bird royal when at large and unmarked, and is consequently afforded protection. Notes: soft and low, plaintive, and of little variety. Local or other name: Common Swan. Male and female Swans are called "Cob" and "Pen" respectively on the Thames. A close sitter.

SWAN CARRYING CYGNETS.

---

## SWIFT.

(*Cypselus apus.*)

Order PICARIÆ; Family CYPSELIDÆ (SWIFTS).

SWIFT ON ROPE.

*Description of Parent Birds.*—Length about eight inches; beak very short, black, and with an extraordinary width of gape. The whole of the plumage is a dingy-black, except the chin, which is of a dirty-white colour. The tail is of medium length and forked, and the wings very long and sword-like; legs, toes, and claws black. The feet have all the four toes in front.

The female is similar to the male in size and colour. Easily distinguished from the Swallow by large size, shorter tail, and colour.

*Situation and Locality.*—Holes in church towers, chimneys, sea cliffs, the walls of old ruins, sometimes within even six feet of the ground, under the tiles of houses and barns. Our illustration is from a photograph of part of Gracious Street, Selborne,

GRACIOUS STREET, SELBORNE.

where astonishing numbers of Swifts nest every year. The tenant of a cottage where Swifts breed once informed me that he could hardly ever sleep after daybreak on account of the noise made by the hungry young birds. Throughout the British Isles; though there are districts where the bird is not met with.

*Materials.*—Scarcely any, consisting only of a few straws lined with feathers, and often glued or cemented together with viscid saliva. The Swift will turn to account the old nest of any other bird, provided it is suitably situated, or even lay its eggs amongst a collection of cobwebs and dust.

*Eggs.*—Two, sometimes three and even four; white, unmarked, and of a narrow elongated shape. Size about 1.0 by .66 in.

*Time.*—May and June.

*Remarks.*—Migratory, arriving in April and May, and leaving in August and September. Note: a harsh scream. Local and other names: Black Martin, Screech Martin, Screecher, Screamer, Squeaker, Deviling. Gregarious, and very fond of its old nesting haunts. Sits closely.

---

## TEAL.

(*Querquedula crecca.*)

Order ANSERES; Family ANATIDÆ (DUCKS).

*Description of Parent Birds.*—Length about fourteen inches and a half; beak of medium length, fairly straight, and almost black. Irides hazel. Head and upper neck chestnut; a narrow line of buff starts from the base of the bill, goes upward, and passes over the eye and ear-coverts, and onward to the back of the head; a second commences in the front corner of the eye, passes under it, and ends behind the ear-coverts. All the feathers between these two lines are of a rich glossy green; back of lower part of neck, scapulars, and upper part of back, waved or barred with narrow transverse black and white lines; lower part of back shaded with dark brown; wings dark brown, beautifully barred with a patch of glossy green and a line of white; upper tail-coverts nearly black, edged with reddish-brown; tail feathers pointed and brown; lower half of neck, in front, pale purplish-white, spotted with black; breast and belly dusky-white; sides and flanks barred with fine wavy lines of black and white; under tail-coverts velvet black; legs, toes, and webs greyish-brown.

The female is much subdued in coloration; her head is light brown, speckled with a darker tint of the same colour; the green spangle on the wing is velvety black; back dark brown, the feathers being edged with a lighter tinge of the same colour; breast and under parts dull white, spotted with

NEWLY HATCHED YOUNG TEAL IN NEST.

dark brown. The male assumes female plumage about the end of July.

*Situation and Locality.*—On the ground amongst heather, rushes, sedges, and coarse grass, near lakes and small sluggish streams, in mountain swamps, by pools and tarns, and in moss bogs. In nearly all suitable districts throughout the British Isles, perhaps scarcest in the south.

*Materials.*—Dried sedges, flags, rushes, reeds, and grass, lined with down from the bird's own body.

*Eggs.*—Eight to fifteen, usually nine or ten; buffish or creamy white, sometimes very faintly tinged with green. Distinguishable from those of

TEAL'S NEST AND EGGS.

the Garganey only by down tufts which are brown without white tips. Size about 1.7 by 1.35 in.

*Time.*—May and June.

*Remarks.*—Resident and partially migratory. Notes: call, *krik;* alarm, *knake.* Local or other name: Common Teal. Sits closely, and when disturbed is wonderfully cunning in trying to decoy the intruder away.

---

## TERN, ARCTIC.

(*Sterna macrura.*)

Order GAVIÆ; Family LARIDÆ (GULLS).

ARCTIC TERN.

*Description of Parent Birds.*—Length about fifteen inches. Bill, rather long, straight, slender, sharp-pointed, and pinky-red. Irides dark brown. Upper part of head and nape black; back, wing-coverts, and wings, French grey; tail-coverts and quills white, with the exception of the two longest feathers, which are grey. The wings are very long, and the tail much forked. Cheeks and chin white; throat and sides of neck ash-grey; breast, belly, and vent, French grey. Legs, toes, and webs, orange; claws black.

The female is similar to the male. This species can be distinguished from the next described by the facts that the bird has no black tip to its bill, has shorter legs, and longer outer tail quills.

*Situation and Locality.*—On the ground, amongst sand, shingle, and on bare rock, near the edge of

ARCTIC TERN'S EGGS.

the water; on low islands, and at suitable places on mainland shores; on the Farne Islands, on the Yorkshire side of the mouth of the Humber; the Scilly Isles; on the Welsh coast, Lancashire, Cumberland; generally round the Scottish coast, and at various suitable places in Ireland.

*Materials.*—None whatever, as a rule, and I am inclined to think that where bits of grass and seaweed are found their presence is accidental.

HOME OF THE ARCTIC TERN.

*Eggs.*—Two or three, varying in ground-colour from pale bluish-green to brownish-buff, blotched and spotted with varying shades of brown and grey. They are found nearer the water's edge as a rule, are slightly smaller, more boldly marked, and inclining to green in the tinge of their ground-colour, than those of the Common Tern. It is, however, generally impossible to distinguish them. Size about 1.55 by 1.1 in. (*See* Plate XII.)

*Time.*—May and June.

*Remarks.*—Migratory, arriving in April and May and departing in September and October. Note: a prolonged *krr-ee*. Local or other name: none, except Sea Swallow, which applies generally to the Terns. Gregarious. A light sitter, the whole colony, when visited, rising and fluttering overhead in a noisy throng. The bird is not shy, however, and will alight after a little while and sit on its eggs within fifteen or twenty yards of the intruder.

---

## TERN, COMMON.

(*Sterna fluviatilis.*)

Order GAVIÆ; Family LARIDÆ (GULLS).

COMMON TERN ON NEST.

*Description of Parent Birds.*—Length about fourteen and a half inches. Bill rather long, slender, straight, sharp-pointed, and pinky red in colour, except at the tip, which is black. This fact and its longer legs distinguish it from the Arctic Tern. Irides dark brown. Upper part of head and nape black; back ash-grey; wings very long, and same colour as the back. Tail much forked and white, except the outer webs of the two longest feathers, which are ash-grey. Chin, throat, breast, belly, vent, and under tail-coverts white, distinguishing the bird from the Arctic Tern, which is grey on its under-parts. Legs, toes, and webs crimson; claws black.

*Situation and Locality.*—A mere hollow on the ground, amongst shingle, sand, coarse grass and vegetation, on rocks and dried wrack; on small islands and quiet stretches of shingly beach round the coasts of the British Isles. Less numerous

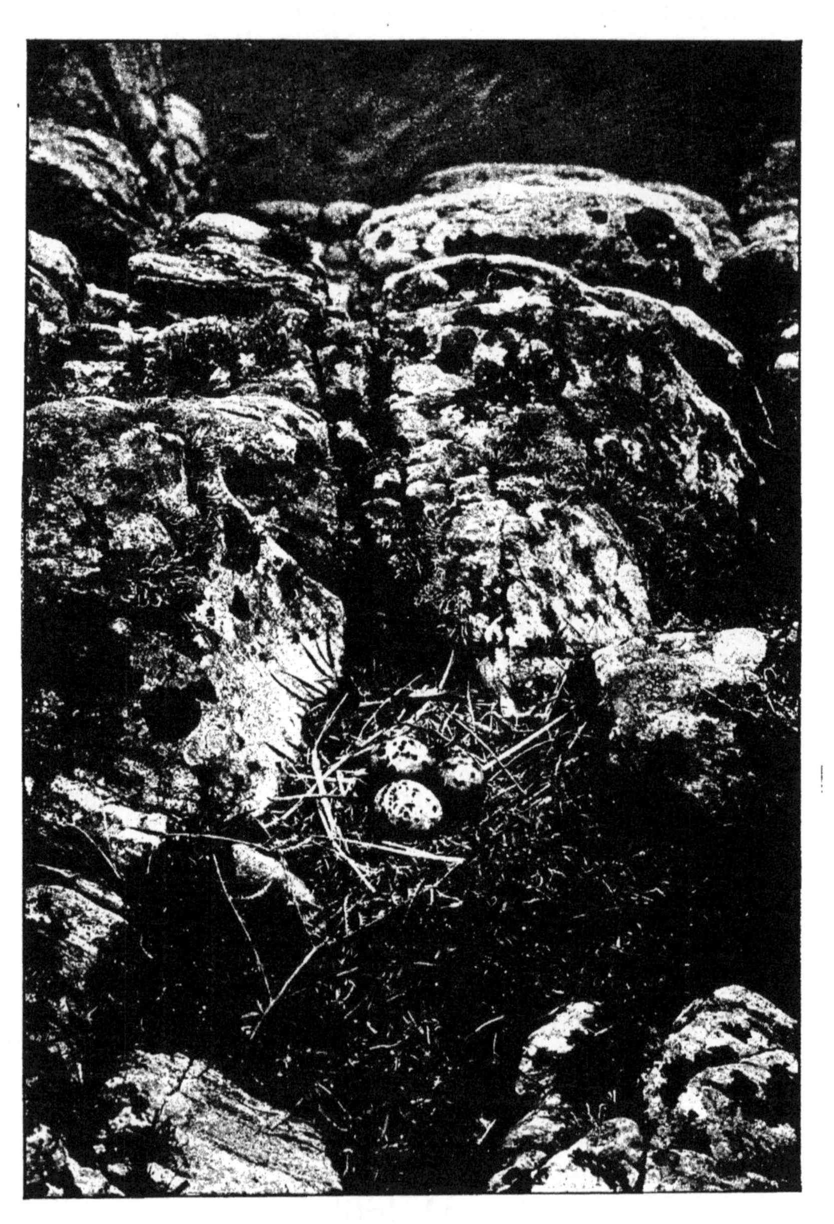

COMMON TERN'S NEST AND EGGS.

round the northern and western coasts and islands of Scotland than the Arctic Tern, but more numerous on the southern and western coasts of England.

*Materials.*—Dry grass, used as a lining to the slight declivity made or chosen. When the eggs are laid on bare rocks, sometimes a slight kind of mat of grass is made. Often there is no kind of material whatever, even when the eggs are laid in this situation.

*Eggs.*—Two or three. Ground-colour light stone or buff to olive- or umber-brown, with ash-grey and light and dark brown spots. Subject to great variation. A trifle larger than those of the Arctic Tern, less boldly marked and lacking green tinge. Size about 1.7 by 1.15 in. (*See* Plate XII.)

*Time.*—May and June.

*Remarks.*—Migratory, arriving in May and departing in August, September, and October. Note: a sharp, angry *pirre.* Local names: Sea Swallow, Tarney or Pictarney, Tarrack, Tarret, Rittock, Rippock, Sporre, Scraye, Pirr, Gull Teaser. Gregarious; sits lightly, and flies overhead when disturbed, uttering its sharp cry.

COMMON TERNS' EGGS ON THE FARNE ISLANDS.

## TERN, LESSER. *See* TERN, LITTLE.

See Tern, Little.

---

## TERN, LITTLE. *Also* LESSER TERN.

(*Sterna minuta.*)

Order GAVIÆ ; Family LARIDÆ (GULLS).

*Description of Parent Birds.*—Length between eight and nine inches. Bill fairly long, straight, and orange-coloured, except at the tip, which is black. Irides dusky. Forehead white, crown and nape deep black. Back and wings French grey, the first two wing-quills being a trifle darker ; the wings are long and narrow. Upper tail-coverts and tail, which is much forked, white. Chin, throat, sides of neck, breast, belly, and vent, clean glossy white. Legs, toes, and membranes orange.

*Situation and Locality.*—On the ground, on sandy, flat coasts interspersed with banks of shells, gravel, and shingle. Some authorities assert that the bird scrapes a slight hollow for the reception of its eggs, whilst others deny this. According to my experience the bird may scrape a slight declivity when breeding on sand, but not when dropping her eggs amongst pebbles. Small colonies are still said to breed on the Kentish side of the mouth of the Thames ; also on the coasts of Essex, Suffolk, Norfolk, Lincolnshire, Yorkshire, Cumberland, and Lancashire, and in suitable places round the Welsh, Scottish, and Irish coasts. It also breeds in a few suitable inland places.

*Materials.*—None, the eggs harmonising well with their surroundings.

*Eggs.*—Two to four, generally two or three, varying in ground-colour from stone-yellow to pale brown, spotted, speckled, and blotched with grey

LESSER TERN'S EGGS.

and dark chestnut-brown. Distinguished by smaller size of eggs and also of parent birds. Size about 1.25 by .95 in. (*See* Plate XII.)

*Time.*—June.

*Remarks.*—Migratory, arriving in May and departing in September or early October. Note: a sharp *pirre*. Local and other names: Lesser Sea Swallow, Lesser Tern. Gregarious. When a colony is visited the birds fly boldly round, uttering their sharp cry, and settle quite close to the intruder.

LESSER TERN ON EGGS.

---

## TERN, ROSEATE.

(*Sterna dougalli.*)

Order GAVIÆ; Family LARIDÆ (GULLS).

*Description of Parent Birds.*—Length about fifteen and a half inches. Bill rather long, straight, and sharp pointed; from the tip to the nostrils it is black, and thence to the base red. Irides dark brown. Crown and back of head black; neck white all round; back and wings ash-grey; tail much forked, long, and pale ash-grey. Breast, belly, sides, vent, and under tail-coverts white, tinged strongly with rose colour. Legs, toes, and

BLACK GUILLEMOT.
(*See p.* 155.)

PUFFIN.
(*See p.* 306.)

BLACK-THROATED DIVER.
(*See p.* 71.)

NOTE.—*In referring to the eggs the above names should be read from*

webs red. It may be distinguished from the other members of the Tern family by its rose-coloured under-parts and elegant, attenuated form. It has been mentioned as occupying the same place amongst Terns as the greyhound does amongst dogs.

The female is very similar to the male.

*Situation and Locality.*—On the ground, in a slight hollow, amongst sand and shingle, on low rocky islands, such as the Farne and Scilly. So far as the British Isles are concerned, this beautiful bird, as a nesting species, was supposed at one time to have become banished. A few pairs, however, attempt to re-establish themselves from time to time in old haunts, but soon dwindle and disappear if their whereabouts become known to collectors. Even at such places as the Farne Islands, where they are afforded protection, the birds do not appear to make much headway.

*Materials.*—Sometimes with, and at others without, a slight lining of bents.

*Eggs.*—Two or three. "Ground colour creamy white or buff-brown, blotched and clouded with bluish-grey and rich brown," according to Mr. Saunders. Subject to same variations as the Common and Arctic Terns, but a trifle longer than the second. Distinguishable, however, only by the appearance of parent birds. Size about 1.7 by 1.15 in. (*See* Plate XII.)

*Time.*—May and June.

*Remarks.*—Migratory, arriving in May and departing, probably about the time of the other Terns. Note: *krr-ee.* Local or other name: none, except Sea Swallow, which, as before stated, is applied to all the Terns. Gregarious, and not a close sitter.

## TERN, SANDWICH.

(*Sterna cantiaca.*)

Order GAVIÆ ; Family LARIDÆ (GULLS).

*Description of Parent Birds.*—Length about fifteen inches ; bill rather long, straight, slender, and pointed ; black in colour, except at the tip, which is primrose-yellow. Irides hazel. Crown of the head and nape black ; the feathers at the back of the head are rather long and form a loose, pointed plume ; back pearl-grey ; wings very long, pearl-grey in colour, except the longest quill-feathers, which are rather darker ; tail white and much forked; chin, throat, breast, belly, and vent white, sometimes tinged with salmon pink ; legs, toes, membranes, and claws black.

The female is very similar to the male.

*Situation and Locality.*—On the ground, in a slight hollow in the sand or pebbles on low rocky islands. The principal colony in this country is on the Farne Islands. Our full-page illustration is from a photograph taken on a slight ridge of sand, measuring about twenty by seven yards. The keepers informed us that in 1892 they counted 210 Sandwich Terns' nests on the same ridge. When our boat neared the Tern Island (Wamseys) the birds (Sandwich, Arctic, and Common) rose in a perfect cloud, and whirled round and round, for all the world like a thick shower of snow played upon by fickle gusts of wind. Small colonies breed at Ravenglass, in Cumberland, and at various points on the Scottish and Irish coasts.

*Materials.*—Generally none ; sometimes, a few bits of dead herbage, and occasionally even a liberal supply of dry grass is used.

SANDWICH TERNS' EGGS AT THE FARNE ISLANDS.

*Eggs.*—Two or three, usually the former number; varying very much in colour, from creamy-white to buffy stone-colour, spotted, blotched, and clouded and scrolled with grey, deep rich brown, and chestnut. During one of my visits to the Farne Islands I saw one of a uniform dark brown, although its fellow egg was of the normal coloration. Size of parent birds and eggs prevents confusion with those of any other Tern. Size about 2.1 by 1.4 in. (*See* Plate XII.)

*Time.*—May and June.

*Remarks.*—Migratory, arriving in April and May and departing in August. Note: a hoarse and grating kind of *kirkitt* or *kirhit*. Local and other names: Great Tern, Sea Swallow. Gregarious, sits lightly, and flies round the intruder, uttering its hoarse cry. I am very pleased to be able to say that this species is extending its breeding area in the British Islands.

SANDWICH TERNS AT THE FARNE ISLANDS.

## THICKNEE. *See* Curlew, Stone.

---

## THRUSH, COMMON. *See* Thrush, Song.

---

## THRUSH, MISSEL. *Also* Mistletoe Thrush.

(*Turdus viscivorus.*)

Order Passeres ; Family Turdidæ (THRUSHES).

MISSEL THRUSH AT NEST.

*Description of Parent Birds.* — Length about eleven inches ; bill moderately long, slightly curved downwards, and dark brown, with a yellowish tinge at the base of the lower mandible. Irides hazel. Top of the head, nape, and back, light brown ; rump yellowish-brown ; wings dark brown, the coverts being tipped and the quills edged with wood-brown ; the tail is darkish brown, the outside feathers edged and tipped with greyish-white ; cheeks, chin, throat, breast, and under-parts straw colour or yellowish-white, lightest on the belly and vent ; the throat and breast are marked with triangular spots of blackish-brown, and the belly with roundish ones of the same colour ; legs and toes light brown ; claws brownish-black.

The female is similar to the male. Easily distinguished from Song Thrush by its larger size, greyer colour of the back, and the fact that the bird conspicuously shows a white quill on either side of the tail when flying away from the observer.

*Situation and Locality.*—Generally near the top of a tree, where the trunk ends and two or three strong branches spring from it, or on a strong bough close to the trunk, at varying heights. I have sometimes found it only three feet from the ground, and at others as many as forty. In orchards, large gardens, woods, plantations, parks, and tree-fringed streams all over the United Kingdom. It is a brave bird, and I have seen it sitting on its eggs when one side of the tree has been plastered white with wind-driven snow.

*Materials.*—Slender twigs, grass stems, wool, moss, and mud, with an inner lining of fine dry grass. The wool often hangs down from the sides of the nest in long conspicuous rags. On one occasion I found, quite by accident, a nest of this species adorned with fresh ivy leaves in such a way as to make it harmonise with the green moss-clad branch upon which it rested. On the other hand, I have seen a nest adorned with large white feathers from a barndoor fowl's wing in such a way as to make the whole structure look most conspicuous.

*Eggs.*—Four to six, according to some authorities. Mr. Dixon, however, says never more than four. Mr. Seebohm says they very rarely exceed four, and in but very few cases are less. Messrs. Dresser and Sharpe say the number is usually five, sometimes four; Waterton says generally five; Macgillivray, usually four, or from three to five. I have certainly heard of more than four; but although I have examined a goodly number of nests, I personally never saw more, except on one occasion near to Stalham in Norfolk. They vary in colour, some being greyish-green with underlying markings of grey, and blotches and spots of reddish-

MISSEL THRUSH'S NEST AND EGGS.

brown. Others are reddish-grey in ground-colour, with brownish-red markings, which vary in size and distribution. Size about 1.3 by .88 in. (*See* Plate IV.)

*Time.*—February, March, April, May, June, and July. I have found them in every month but the last.

*Remarks.*—Resident, but subject to southern movement in winter. Song loud and defiant, but not considered of much value by bird-fanciers, as it is somewhat melancholy and made up of five or six broken strains; alarm note, a jarring kind of scream. Local and other names: Holm Thrush, Storm Cock, Holm Screech, Mistletoe Thrush, Missel Bird, Bell Throstle, Screech Thrust. Sits pretty closely, and makes a great deal of demonstration when disturbed.

A brave Missel Thrush has been known to attack a weasel in defence of its young, and perish for its temerity. I once saw a bird of this species dash at a stuffed owl fixed in a tree near her nest.

YOUNG MISSEL THRUSHES.

## THRUSH, SONG. *Also* Thrush *or* Common Thrush.

(*Turdus musicus.*)

Order Passeres ; Family Turdidæ (THRUSHES).

SONG THRUSH.

*Description of Parent Birds.*—Length about eight and a half inches. Bill of medium length, nearly straight, and dusky. Irides hazel. Head, nape, back, wings, rump, tail-coverts, and quills yellowish-brown, spotted with darker brown on the sides of the head, and edged with a lighter tinge on the wing-quills. Throat, breast, and under-parts pale tawny-yellow; lighter on the vent and under tail-coverts. The space from the throat to the thighs is studded with arrowhead-like spots. Legs and toes pale brown; claws darker.

The female is a trifle smaller than the male, and the spots on her breast are larger and the ground colour lighter.

*Situation and Locality.*—In evergreens, especially early in the spring, hedgerows, bushes, in ivy growing against walls and trees, in holes and on "throughs" of dry walls; on ledges of rock, on beams and in holes of barns, and sometimes quite on the ground; in woods, plantations, on commons, hedges, trees and bushes growing by the side of brooks. Throughout the British Isles, with few exceptions, and those where no cover is afforded.

*Materials.*—Twigs, coarse grass, moss, and clay, with an inner lining of cow-dung or mud, mixed with

SONG THRUSH'S NEST AND EGGS.

bits of decaying wood; sometimes thickly studded with bits of rotten wood. During droughty weather the bird is occasionally unable to secure any suitable materials wherefrom to make an inner lining, and has in consequence to make a nest very similar to that of the Common Blackbird.

*Eggs.*—Four to six, of a beautiful deep greenish-blue, spotted with black. The spots sometimes describe a well-defined ring round the larger end,

FEMALE SONG THRUSH AT NEST WITH YOUNG.

at others they are sparingly scattered over the egg, and in rare cases are absent altogether. Very variable in size. Average measurements about 1.05 by .8 in. (*See* Plate IV.)

*Time.*—February, March, April, May, June, and July; sometimes as late as August, and even October. The bird may sometimes be seen bravely covering her eggs when the ground has been thickly mantled in snow.

*Remarks.*—Resident, subject to local movement, and partially migratory. Notes: call, *sik, sik, sik, sik, siki, tsak, tsak.* The song of the cock is well known and highly esteemed. In North Yorkshire it is verbalised as *Pay thy debt, pay thy debt, skitting Mick, skitting Mick.* Local and other names: Throstle, Mavis. Sits closely, and protests loudly against molestation.

---

## TIT, BEARDED.

(*Panurus biarmicus.*)

Order PASSERES; Family PANURIDÆ (REEDLINGS).

MALE BEARDED TIT.

*Description of Parent Birds.*—Length just over six inches, the tail forming about half of this; bill short, upper mandibles slightly curved downwards, and yellow. Irides yellow. Crown bluish-grey; nape, shoulders, back, and rump, golden brown; wings black and greyish, the feathers being bordered and tipped with white and deep rusty-red; upper tail-coverts and tail, which is graduated and wedge-shaped, deep rufous brown, some of the outer feathers being tipped and edged with white and greyish-white. A black patch extends from the base of the bill to half way over the eye and, passing downwards, ends in a tuft of elongated, tapering black feathers growing from the side of the chin and throat very like a moustache. Centre of chin and throat dirty-white; breast flesh-colour; belly and vent like the back, but brighter; under tail-coverts black; legs, toes, and claws black.

The female has the crown dull rusty-brown;

BEARDED TIT'S NEST AND EGGS.

tufts on the sides of the chin pale brownish-fawn; chin and throat mixed with light brown; breast of a lighter tinge than in the male; and under tail-coverts pale golden-brown.

*Situation and Locality.*—Close to the ground, in a tuft of coarse grass, or sedge; sometimes amongst broken-down reeds, but never suspended between stems of any kind. It is well hidden. This interesting and rare species has its headquarters amongst the extensive reed-beds and marshes round the Norfolk Broads. It also breeds in one or two places in Suffolk, and has been seen within twenty miles of London within the last few years; but, of course, the exact localities must, for obvious reasons, remain nameless, as the bird and its eggs are persistently sought after by collectors, one of whom recently left Broadland with no less than four clutches in his possession. The significance of a raid of this character can be appreciated, when one considers that a competent authority recently gave it as his opinion that the Bearded Tits did not number forty pairs in the whole county.

*Materials.*—Dead aquatic vegetation, such as leaves of reeds and blades of sedges, lined with fine grass and seed-down. It is cup-shaped.

*Eggs.*—Four to seven, white, faintly tinged with cream colour, and marked with small specks, short irregular streaks and splashes of dark brown. Distinguished by situation, size of eggs, and streaky markings. Size about .7 by .56 in. (*See* Plate III.)

*Time.*—March, April, May, June, and July.

*Remarks.*—Resident, but subject to local movement. Notes: shrill and musical when alarmed. They also utter a clear silvery note—which may be very successfully imitated by poising a half-crown on the tip of the finger of one hand and tapping it with a similar coin held in the other—before alighting. Local or other name: Reed Pheasant. Sits closely.

THE HOME OF THE BEARDED TIT.

---

## TIT, BLUE.

(*Parus cæruleus.*)

Order PASSERES; Family PARIDÆ (TITMICE).

BLUE TIT.

*Description of Parent Birds.*—Length from four to four and a half inches. Bill short, strong, and dusky. Irides dark brown. Crown clear blue, under which runs a band of white on either side. From the base of the beak, through the eyes, passes a bluish-black line. Cheeks white; a broadish circle of dusky blue runs round from the back of the head to the throat, where it becomes almost black. Back and rump lemonish-green; wings and tail blue, the former marked with white on the coverts and tertials. Chin and throat blue-black; breast, belly, and under-parts yellow. Legs, toes, and claws dull leaden-blue.

The female is less brilliant and distinctive in her coloration.

*Situation and Locality.*—In holes, in trees, walls, banks, and often in such queer places as disused pumps, letter-boxes, stone bottles, flower-pots, boxes, and cocoanuts, hung in trees for its accommodation. Our illustration is from a photograph of a nest in a hollow fruit tree. The entrance to the nest was in the centre of a decayed branch which had been sawn or broken off close to the trunk, at the place where the dark excrescence-like growth appears, the hole through which the eggs are to be seen being cut artificially through the wood, so as to show its exact position. In barns, stables, cottages, orchards, gardens, woods, and cultivated districts generally, throughout the United Kingdom, with the exception of the islands lying to the west and north of Scotland.

*Materials.*—Grass, moss, hair, and wool; sometimes a few soft leaves woven together, with an inner lining of feathers. I have met with specimens containing few or none of the last.

*Eggs.*—Six to nine, sometimes as many as eleven or twelve, white, spotted with light red or red-brown, sometimes evenly distributed, at others most numerous at the larger end. A sight of parent birds only will definitely settle identification. Size about .6 by .46 in. (*See* Plate III.)

*Time.*—April, May, and June.

BLUE TIT'S NEST AND EGGS.

*Remarks.*—Resident. Note: a peculiar *twe-twe.* Local and other names: Tomtit, Blue Tomtit, Billy Biter or Willow Biter, Blue Bonnet, Blue Cap, Blue Mope, Hickwall, Nun, Titmal. A close sitter, hissing like a snake when disturbed.

---

## TIT, COAL.

(*Parus britannicus.*)

Order PASSERES; Family PARIDÆ (TITMICE).

COAL TIT.

*Description of Parent Birds.*—Length about four and a half inches. Bill short, straight, pointed, and black. Irides hazel. Head, neck, and upper part of breast black; cheeks and nape white. Back, wing-coverts, rump, and tail greyish-blue, with a buffish tinge on the rump. Wing-quills brownish-grey, bordered with green. Lower breast dull white; belly, flanks, vent, and under tail-coverts fawn colour, slightly tinged with green. Legs, toes, and claws black.

The female closely resembles the male. This bird is easily distinguished from the Marsh Tit by means of the white patch on the back of its head and neck.

*Situation and Locality.*—In holes, from three or four to sixteen or eighteen inches deep, in trees, walls, and banks; those in the last-named situations having originally belonged to rats, mice, or moles. The bird will, however, enlarge any selected hole for its accommodation, if necessary. It may be found in orchards, spinneys, coppices, woods, and plantations throughout England, Wales, Scotland, and Ireland.

*Materials.*—Dry grass, moss, wool, and hair, lined liberally with feathers.

*Eggs.* — Five to ten, generally seven or eight, white, spotted and freckled with light red, the markings being generally most numerous at the larger end. They bear a very close resemblance to those of other members of the Tit family, but a sight of the parent birds will readily distinguish them. Size about .62 by .47 in. (*See* Plate III.)

COAL TIT'S NEST AND EGGS
*(Bricks removed to show it.)*

*Time.*—April, May, and June.

*Remarks.*—Resident. Note: a harsh, shrill, *che-chee, che-chee.* Local and other names: Colemouse, Coal Titmouse, Coalhead. Sits closely, and hisses when molested.

## TIT, CRESTED.

(*Parus cristatus.*)

Order PASSERES ; Family PARIDÆ (TITMICE).

CRESTED TIT.

*Description of Parent Birds.*—Length about four and a half inches; bill short, straight, and almost black. Irides hazel. The feathers on the top of the head, especially behind, are lengthened, and form a conspicuous crest; these feathers are dull black in colour, tipped with light grey; back, wings, rump, and tail brown, the quills being somewhat darker. A black streak runs from the base of the beak to the eye, and passes onward between the base of the crest and the ear-coverts to the nape, whence a broader black, curving line descends behind the cheeks and ends abruptly on the sides of the neck. Cheeks white, spotted with black. The black line just described is followed beneath by a broader band of white, which in turn gives place to a narrower curving black line, descending from the back of the head, and, passing in front of the point or shoulder of the wing, joins the black on the upper breast; chin, throat, and upper breast black; lower breast, belly, and flanks dull white, suffused with buff on the sides; under tail-coverts dull buff; legs, toes, and claws lead-grey.

The female resembles the male, except that her crest is shorter, and the black on her chin, throat, and upper breast occupies less space.

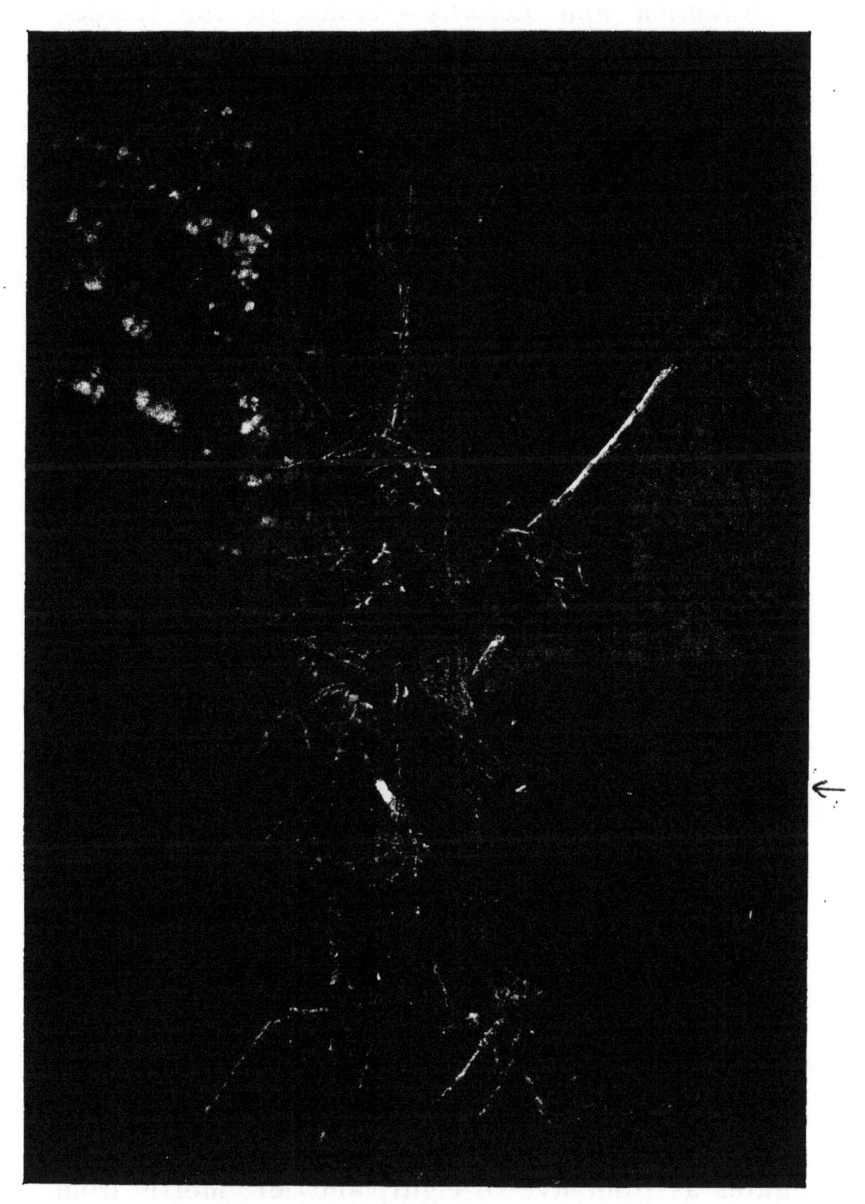

CRESTED TIT'S NESTING HOLE IN DEAD TREE.

*Situation and Locality.*—Holes in the trunks, branches, old stumps of trees, and posts. The hole is situated from a few inches to ten or twelve feet from the ground, and is sometimes dug by the bird's own exertions. On the Continent it patronises the deserted nests of Crows and old squirrel dreys. It breeds in old pine, fir, and oak forests in Ross, Banff, Perth, Inverness, and possibly one or two other favourite localities in Scotland only.

THE HOME OF THE CRESTED TIT.

*Materials.*—Dead grass, moss, feathers, and down of hares and rabbits. The four nests I have examined of this rare and interesting species consisted almost entirely of deers' hair and rabbits' down.

*Eggs.*—Five. Professor Newton says that they do not seem to exceed five in number, which my own experience confirms. Mr. Dixon gives the figures as from five to eight, and Mr. Morris from seven to ten. Possibly the last is the result of some Continental information, as it is said the bird lays from eight to ten eggs there. White, spotted, blotched, and speckled with red of various shades; the spots are generally most numerous round the larger end. A sight of the parent birds and locality of the nest are the only safe means of identification. Size about .65 by .51 in. (*See* Plate III.)

*Time.*—April, May, and June.

*Remarks.*—Resident and very local. Notes: *si, si, si,* followed by a spluttering note like *ptur, re, re, re, ree.* Local and other names: Crested Titmouse. Sits closely.

---

## TIT, GREAT.

(*Parus major.*)

Order PASSERES; Family PARIDÆ (TITMICE).

GREAT TIT.

*Description of Parent Birds.*—Length about five and three-quarter inches; bill of medium length, nearly straight, and black. Irides dusky. Head black; cheeks white; nape greenish-yellow, surrounding a whitish spot; back olive-green; rump bluish-grey; wing-coverts bluish, the larger being tipped with white; quills dusky, edged with light greenish-blue; tail quills dusky, the outer feathers on each side being edged with white; breast and belly yellow, tinged with green and divided longitudinally by a broad black stripe, which commences at the chin and, passing down the throat, is joined by the black on the sides of the neck, and extends to the vent, which is white on either side; lower tail-coverts whitish; legs, toes, and claws bluish-grey.

The female is not so distinctive in coloration; the black stripe on her under-parts is not so wide or pronounced, and vanishes about the middle of the belly. She is also said by some ornithologists to be smaller, but this is certainly not a very noticeable feature.

*Situation and Locality.*—In holes in trees, walls, posts, banks, and buildings. The bird will sometimes excavate a hole for itself in a rotten tree. It also occasionally fixes its abode beneath the nest of a Crow, Rook, or Magpie; inside upturned plant pots standing in gardens, old pumps, and other odd situations. I once found one built inside the old nest of a Blackbird situated in a dense, tangled hedge. Great Tits often return year after year to the same hole to breed. In orchards, gardens, yards, well-timbered commons, woods, plantations, parks, and other wooded places throughout the United Kingdom, with few exceptions.

*Materials.*—Dry grass, moss, hair, wool, and feathers, somewhat carelessly put together. Sometimes no materials at all. Montagu and Morris supposed that where materials were dispensed with altogether, the bird had had her first eggs and nest taken, and had not had time for more nest-building. This theory appears to be perfectly sound, for a few years ago I came across an instance of a bird that had had her first clutch of eggs taken, along with the nest, from a hole in an apple tree, laying again promptly on the rotten powdered wood at the bottom of the hole.

*Eggs.*—Five to twelve, generally seven or eight; a Surrey gardener told me one day that a pair of Great Tits nested in a disused beehive close to his master's house, and that he took the eggs away as

GREAT TIT'S NEST AND EGGS UNDER AN UPTURNED PLANT POT.

laid with the exception of one. Before the bird gave up the task as hopeless she deposited no less than thirty-two eggs inside the old hive. White in ground-colour, spotted and freckled with pale red. The markings vary in size, number, intensity, and position. Sometimes they are generally distributed, at others they form a zone at the larger end of the egg. They are larger than the eggs of any other Tit, but are likely to be confused with those of the Nuthatch if care is not taken. Size about .7 by .53 in. (*See* Plate III.)

*Time.*—April, May, and June.

*Remarks.*—Resident. Notes: a clear, pealing *pinker, pinker,* repeated several times. Local and other names: Oxeye, Pickcheese, Great Black-headed Tomtit, Blackcap, Great Titmouse, Bee-biter, Tom Collier, "Sit-ye-down." Sits closely, and hisses like a snake when disturbed. This bird is somewhat like the Robin in its love for odd nesting quarters. Last spring I had three nests shown to me beneath upturned plant pots standing in gardens, although there were plenty of eligible nesting sites in the immediate neighbourhood.

NEWLY FLEDGED GREAT TITS.

## TIT, LONG-TAILED.

(*Acredula rosea.*)

Order PASSERES ; Family PARIDÆ (TITMICE).

YOUNG LONG-TAILED TIT.

*Description of Parent Birds.*—Length about five and a half inches, of which the bird's abnormally long tail forms a considerable part ; bill very short, and black. Irides hazel. Forehead and crown white. A black streak of variable width commences near the base of the bill, and, passing over the eye and ear-coverts, meets on the back of the neck, and descending forms a kind of triangle, the lowest point of which reaches the middle of the back. The scapulars and rump are suffused with a dull purplish-red ; wing-coverts and quills black, the inner ones of the latter edged with white ; upper tail-coverts and six centre quills black, the remainder of the feathers on either side of the tail being more or less white. The feathers of the tail are much graduated in length, those in the centre being about an inch and three-quarters longer than those on the sides. Cheeks and ear-coverts white, the latter mixed with grey. Chin, throat, breast, belly, and under-parts greyish-white, tinged with purplish-red on the sides, flanks, vent, and under tail-coverts ; legs, toes, and claws almost black.

The female is somewhat similar, with the exception of her head, which has more black upon it ;

however, both sexes are subject to variation in the intensity of coloration.

*Situation and Locality.*—In whitethorn hedges, sloe, gorse, and wild rose bushes. I have seen specimens within a couple of feet of the ground on some occasions, and on others between thirty and forty feet high fixed to the side of a single bough in an oak tree. It is found in nearly all suitable localities throughout the British Isles.

LONG-TAILED TIT AT NEST.

*Materials.*—Moss, lichens, wool, spiders' webs, cunningly felted together, and skilfully formed into an oval-shaped nest, which is plentifully lined with feathers and securely fastened to its surroundings. During the spring of 1905 I one day found a nest which had been shown me by a woodcutter torn out and destroyed, and even amongst the fragments of the structure I counted over three hundred feathers which had belonged to seven different birds. Upwards of 2,000 have, however, been found in one nest.

*Eggs.*—Seven to ten; as many as twenty have been found, but such a number was undoubtedly the production of two hens. White or rosy-white until blown, by reason of the yolk showing through the delicate shell, with very small, faint red or

LONG-TAILED TIT'S NEST AND YOUNG.

reddish-brown spots, sometimes collected round the larger end, at others sparingly scattered over the entire surface ; occasionally without spots altogether. The smaller number of spots, the character and situation of the nest, and the appearance of the parent birds readily identify the eggs of this Tit. Size about .57 by .44 in. (*See* Plate III.)

*Time.*—March, April, May, and June.

*Remarks.*—Resident. Notes: a sharp chirrup or twitter, varied by a lower and hoarser note. Local and other names: Mufflin, Poke Pudding, Longpod, Bottle Tit, Oven-builder, Caper Long-tail, Long-tail Pie, Long-tailed Capon, Bottle Tom, Mum Ruffin, Long-tailed Mag, Huckmuck (a name also applied to the Willow Warbler), Long-tailed Mufflin, etc. Sits closely, with the tip of her tail protruding from the hole. The male keeps the female company in the nest at night time during the period of brooding.

---

## TIT, MARSH.

(*Parus palustris.*)

Order PASSERES ; Family PARIDÆ (TITMICE).

*Description of Parent Birds.*—Length about four and a half inches. Bill short, straight, sharp-pointed, and black. Irides dark hazel. Forehead, crown, and nape deep black. Back, wing-coverts, and upper tail-coverts ashy-brown, mixed with a greenish tint. Wing and tail-quills greyish-brown, with edges of a lighter tinge. Cheeks dirty-white ; chin black ; throat and breast dull greyish-white ; belly and vent of the same colour, tinged with brown. Legs, toes, and claws bluish-black.

MARSH TITS AND NESTING SITE.

The female is similar in appearance to the male. The bird may be easily distinguished from the Coal Tit by its having no white on the back of the head and neck or wing-coverts.

*Situation and Locality.*—Holes in trees, preferably pollards, gate-posts, walls, and banks, at no great height from the ground. Instances are on record of rabbit-burrows and rat-holes doing duty as nesting sites. In orchards, woods, by the side of sluggish rivers, and in hedgerows of cultivated districts. It is met with in most parts of England suitable to its habits. Scotland and Ireland can both claim it, but it is somewhat scarce, especially in the northern parts of both countries.

*Materials.*—Moss and fine dried grass, lined with wool, feathers, hair, rabbits' down, ripe catkins of the willow, the whole being compactly knitted together, and tightly wedged into the situation chosen for their reception.

*Eggs.*—Six to ten, white, spotted with reddish-brown, more thickly at the larger end. The spots are variable in size, number, and distribution. They very closely resemble those of the Tree Creeper, Blue Tit, Coal Tit, Nuthatch, and Great Tit, although, as a rule, they are somewhat smaller than the last. Size about .63 by .49 in. (*See* Plate III.)

*Time.*—April, May, and June.

*Remarks.*—Resident. Notes: call, *chee-chee* or *peh, peh,* uttered quickly, and several times in succession, and a kind of whistle, made use of only in the spring, according to Montagu. Local and other names: Black Cap, Little Black-headed Tomtit, Willow Biter, Coalhead. Sits closely, and hisses and bites when disturbed.

(*See p.* 274.)

(*See p* 231.)

RAZORBILL.
(*See p.* 317.)

CORN CRAKE.
(*See p.* 42.)

COOT.
(*See p.* 36.)

SPOT
(S

NOTE.—*In referring to the eggs the above names should be read from left*

PLATE 14

## TWITE. *Also* MOUNTAIN LINNET.

(*Linota flavirostris.*)

Order PASSERES; Family FRINGILLIDÆ (FINCHES).

TWITE ON NEST.

*Description of Parent Birds.*—Length about five and a quarter inches; bill short, broad at the base, and pale yellowish flesh colour. Irides hazel. Crown, neck, back, and upper tail-coverts dark brown, the feathers being edged with light rufous-brown; rump purplish-red; wing and tail quills very dark brown, more or less edged on the outer webs with white. The feathers round the base of the beak and below the eyes tile-red; sides of the head dark brown, edged with a lighter tinge; chin and throat rufous, lighter on the breast and sides, which are speckled with brown; belly nearly white; vent tinged with brown; under tail-coverts almost white; legs, toes, and claws dusky; tail slightly forked.

The female is lighter coloured on her upper parts, and lacks the red on her rump.

*Situation and Locality.*—Amongst tall heather, ling, brushwood, and furze, in honeysuckle, small gooseberry bushes, in large holes of old dry stone walls, under loose rocks. I once saw one built in a creeping geranium inside a small Hebridean greenhouse. Frequently on the ground against the side of a bank or by a stone in moorland districts in the north of England, Scotland and the surrounding

TWITE'S NEST AND EGGS.

islands, and Ireland. According to my experience it is more numerous as a breeding species in the Outer Hebrides than anywhere else in our country.

*Materials.*—Twigs, fibrous roots, grass stalks and blades, moss, and wool, with an inner lining of feathers, hair, or down.

*Eggs.*—Four to seven, generally five or six; very similar, indeed, to those of the Linnet. Pale bluish-green, spotted and streaked with reddish-brown and dark brown; sometimes streaked with the lighter reddish tinge. Some authorities say that they are a little more streaked, and that the light red markings are less frequent than in those of the Linnet. The markings are generally most numerous on the larger end of the egg. Easily distinguished by the appearance of the parent birds. Size about .69 by .5 in. (*See* Plate II.)

*Time.*—May and June.

*Remarks.*—Resident in its breeding haunts, but a winter visitor to the more southern portions of England. Notes: *twite;* the cock has a pleasing little song. Local and other names: Mountain Linnet, Twite, Finch, Heather Lintie. Sits closely.

YOUNG TWITES.

## WAGTAIL, BLUE-HEADED.

(*Motacilla flava.*)

Order Passeres; Family Motacillidæ (WAGTAILS).

*Description of Parent Birds.*—Length about six and a half inches; bill fairly long, slender, straight, and black. Irides dullish brown. Crown and nape bluish-grey; scapulars, back, and upper tail-coverts greenish-olive, suffused with yellow; wing-coverts and primaries dark brown, the former, as well as the tertials, bordered with greyish-yellow. The tail is black in the centre and white on the outer edges. Over the eye and ear-coverts is a white streak, also one of shorter dimensions under the eye; ear-coverts bluish-grey; chin and cheeks white; throat, breast, belly, vent, and under tail-coverts golden-yellow; legs, toes, and claws black.

The female is somewhat smaller and less brilliant and distinctive in coloration. The bird may be distinguished from the Yellow Wagtail, which it closely resembles, by its bluish-grey head and the white streak over the eye and ear-coverts.

*Situation and Locality.*—On the ground, amongst meadow grass, on hedgerow banks, amongst the exposed roots of trees, in pastures, grass meadows, and cornfields, according to Continental observations.

*Materials.*—Dead grass, moss, and fibrous roots, lined with horsehair.

*Eggs.*—Four to six, usually five; quite similar to those of the Yellow Wagtail; greyish-white, suffused, mottled, or spotted with varying shades of brown; sometimes marbled with a few fine lines of dark brown. Size about .78 by .56 in. (*See* Plate III.)

*Time.*—May and June, according to Messrs. Dixon and Miller Christy; but these months are either based upon Continental observations or deductions from the laying season of the Yellow Wagtail.

*Remarks.*—About forty specimens of the Blue-headed Wagtail have been procured in this country, shot in January, April, June, and October. Its nest has only been met with on one or two occasions at Gateshead, but it is thought that from its close similarity to the Yellow Wagtail it has often been overlooked. Call note: *chit-up.* Local or other name: Grey-headed Wagtail. There is, so far as I can gather, no precise information forthcoming as to whether the bird is a close sitter or not.

---

## WAGTAIL, GREY.

(*Motacilla melanope.*)

Order Passeres; Family Motacillidæ (WAGTAILS).

GREY WAGTAIL ON NEST.

*Description of Parent Birds.*—Length about seven and three-quarter inches, nearly half of which is accounted for by its unusually long tail. Bill of medium length, nearly straight, and dusky-brown. Irides dark hazel. Crown and sides of head bluish-grey; a narrow white streak runs over the eye and ear-coverts. Back of neck, back, scapulars, and rump bluish-grey; wing-coverts black, or very nearly so, tipped with buffish-white; quills black, some of the inner ones edged on the outer webs with yellowish-white, and liberally marked on the inner, towards the base, with white. Upper tail-coverts greenish-yellow; tail black, yellowish on the edges of the centre feathers towards the base; the two outside quills on either side white, with the exception of a narrow black line on the outer web of the second feather. Chin and throat black, separated from the sides of the head and neck by a white line; breast, belly, and under-parts bright yellow. Legs, toes, and claws pale brown.

*Situation and Locality.*—On shelves of rock, in crevices, in rough, rocky, and uneven banks, holes in stone walls, behind or under large stones, rarely far away from water. It is very local, and, like the Dipper, seems to lay claim to a certain length of stream. I am familiar with two waterfalls on moorland becks in the north of England where a Dipper and a Grey Wagtail nest almost yearly within a few yards of each other. I have known the bird on one occasion become foster-parent to a young Cuckoo. It breeds in the western and northern counties of England; in Wales, Scotland, and in parts of Ireland. Our illustrations were procured in Westmorland.

*Materials.*—Rootlets, grass, and moss, lined with horse and cowhair; sometimes a few feathers.

*Eggs.*—Four or five, occasionally six, of a greyish-white ground-colour, spotted and speckled with pale brown. Sometimes the ground-colour is buffish

GREY WAGTAIL'S NEST AND EGGS.

and the markings creamy-brown. Occasionally a few streaks of dark brown are present. Much like the eggs of the Yellow and Blue Headed Wagtails, also Sedge Warbler, but easily identified by locality, situation, and a sight of the parent birds. Size about .75 by .56 in. (*See* Plate III.)

*Time.*—April, May, and June.

*Remarks.*—Resident, but subject to local migration. Notes: *sziszi* or *zisy*, sharply uttered. Local and other names: Dun Wagtail, Nanny Washtail, Grey Wagster. Sits closely, and when disturbed hovers round with her mate, uttering a note of alarm.

---

## WAGTAIL, PIED.

(*Motacilla lugubris.*)

Order PASSERES; Family MOTACILLIDÆ (WAGTAILS).

PIED WAGTAIL.

*Description of Parent Birds.*—Length about seven and a half inches, several of which are accounted for by the somewhat abnormally long tail. Bill moderately long, nearly straight, slender, and black. Irides dusky. Forehead, sides of head, round the eyes, and a portion of the sides of the neck, white. Latter half of crown, nape, back, and upper tail-coverts black; wing-coverts black, edged and tipped with white; quills black, some of them bordered with greyish-white

PIED WAGTAIL'S NEST AND EGGS.

and white; tail-quills black, except the two outside feathers, which are nearly all white. Chin, throat, sides, and flanks black; breast, belly, and under-parts white. Legs, toes, and claws black.

The female is somewhat smaller, and dusky-grey on the back, where the male is black.

*Situation and Locality.*—In ivy, growing against walls and trees, in holes in dry walls, bridges, niches of rock, on ledges, and tufts of grass growing from crevices of rock; in faggot, hay, and brick stacks, collections of rough boulders, and numerous other situations, generally near fresh water, throughout the British Isles. It has been recorded in such curious situations as a potato top, and under a railway switch.

*Materials.*—Dry grass, roots, and moss; sometimes a few dead leaves or fern-fronds, with an inner lining of wool, feathers, horsehair, cowhair, and rabbit down. The materials vary both in quantity and character, according to situation.

*Eggs.*—Four to six, greyish-white, thickly speckled with ash-grey or light brown. They vary a good deal, according to the tint of the ground-colour, size of the spots, and their colour. Those in the nest represented on page 449 were of a bluish-white, marked all over with small ash-grey spots. Indistinguishable from those of the White Wagtail or certain varieties of the House Sparrow, except by the parent birds in the former case and the nest and its position in the latter. Size about .8 by .6 in. (*See* Plate III.)

*Time.*—March, April, May, and June.

*Remarks.*—Partially resident, but with a southern winter movement. Notes: *chiz-zit, chiz-zit.* Local and other names: Dishwasher, Black and White Wagtail, Washtail, Nanny Washtail, Wagster, Water Wagtail, Washerwoman. Sits rather closely.

YOUNG WAGTAILS IN NEST.

---

## WAGTAIL, WHITE.

(*Motacilla alba.*)

Order PASSERES; Family MOTACILLIDÆ (WAGTAILS).

*Description of Parent Birds.*—Length nearly eight inches. Bill of medium length, straight, and black. Crown and nape black; back, scapulars, and upper tail-coverts French or light ash-grey. Wings brownish-black, each feather having a broad outer margin of greyish-white, tail-quills black, with the exception of the two centre feathers, which are margined with white, and the two outer feathers on each side, which are white, with black inner webs. The front and sides of the head, together with a patch on either side of the neck, are white. Chin, throat, and upper part of the breast, black. Lower breast, belly, vent, and under tail-coverts white. Legs, toes, and claws black.

The female is less distinctive in coloration. Her

forehead and cheeks are not so pure a white; throat mottled with white; black on back of head occupies less space, and her back is tinged with olive.

*Situation and Locality.*—Similar in all respects to those of the Pied Wagtail.

*Materials.*—Same as employed by last species.

*Eggs.*—Five to seven, of a wider colour variation than those of the Pied Wagtail, according to Mr. Dixon. The ground-colour varies from pure white to bluish-white, speckled all over with different shades of grey and brown; sometimes a few hair-like lines occur at the larger end. The markings vary, both in regard to size and distribution, and there can hardly be any safe means of identification apart from the difference in the parent birds. Size about .8 by .6 in. (*See* Plate III.)

*Time.*—April, May, and June.

*Remarks.*—Migratory, but little is known as to its comings and goings. Although a common bird on the Continent, only a few well-authenticated instances of its breeding in the British Isles are on record, and those in the southern counties of England. It is, however, thought that it may often have been overlooked from the fact that its general appearance and eggs are so similar to those of the Pied Wagtail, to all except the practical and experienced ornithologist. Notes: call, *chiz-zit*. Local and other names: Grey and White Wagtail. The male differs from the Pied Wagtail in being grey on his upper-parts below the nape instead of black, but the females of the two species only differ in that of the White Wagtail being "pearl-grey or very light ash-grey tinged with olive," and that of the Pied Wagtail being "lead-grey mottled with darker feathers" on those parts.

## WAGTAIL, YELLOW.

(*Motacilla raii.*)

Order PASSERES; Family MOTACILLIDÆ (WAGTAILS).

YELLOW WAGTAIL AT NEST.

*Description of Parent Birds.*—Length about six and a half inches. Bill moderately long, straight, slender, and black. Irides hazel. Crown, nape, back, and scapulars light olive. Wing-coverts and primaries darkish-brown, the first-named being tipped, and the tertials bordered and tipped with greyish-yellow. Upper tail-coverts olive; tail-quills brownish-black, with the exception of the two outer feathers, which are white, streaked with black on the inner web. Over the eye and ear-coverts is a line of golden-yellow. Chin, throat, breast, belly, and vent a bright golden-yellow. Legs, toes, and claws black.

The female is much less handsome, her head and back being darker, and the yellow of her breast and under-parts not nearly so bright.

*Situation and Locality.*—On the ground, in the shelter of a tuft of grass, heather, or coarse herbage; sometimes behind the long grass of an overhanging bank, well hidden. I know several places in the north of England where pairs breed year after year with unbroken regularity. In grass meadows, pastures, commons, and other suitable places, pretty generally throughout England, except Cornwall and Devonshire. I have met with it more numerously

YELLOW WAGTAIL'S NEST AND EGGS.

in Norfolk, parts of Suffolk, and in Yorkshire and Westmorland, than anywhere else in the British Isles; also in the south of Scotland, and, to a very limited extent, in Ireland. The bird is wary, and the nest difficult to find. I have watched a pair for three or four hours through my binoculars, and when able to locate the nest pretty closely have still failed to find it.

*Materials.*—Dead grass, fibrous roots, and moss,

YELLOW WAGTAIL'S NEST AND YOUNG.

with an inner lining of horse or cowhair, feathers, or down.

*Eggs.*—Four to six, generally five, ground-colour greyish-white, mottled and spotted with varying shades of brown; sometimes marbled with blackish-brown at the larger end. The markings are thickly distributed over the surface of the egg. They are very similar to those of the Blue-headed Wagtail, Pied Wagtail, and Sedge Warbler. The difference pointed out in regard to the plumage of the first in describing it, and the situation of the two latter, ought to prevent confusion. Size about .78 by .58 in. (*See* Plate III.)

*Time.*—Some very good authorities say April,

but I have never met with a nest so early. May, June, and July.

*Remarks.*—Migratory, arriving in March or April, and leaving in September. Notes: *tzee-tzee, sipp-sipp.* Local and other names: Cowbird, Ray's Wagtail, Yellow Wagster. Sits lightly, and, although by no means shy, is very wary.

---

## WARBLER, DARTFORD

(*Melizophilus undatus.*)

Order PASSERES; Family SYLVIIDÆ (WARBLERS).

DARTFORD WARBLER.

*Description of Parent Birds.*—Length about five inches, nearly half of which is accounted for by the bird's exceptionally long tail; bill fairly long, slightly curved downward, and blackish, with the exception of the base of the lower mandible and along the edges of the upper, which are orange. Irides light or dark red, according to age. Head, neck, back, and upper-tail coverts greyish-black; wings blackish-brown, the quill-feathers being bordered with a lighter tinge; tail blackish-brown, the external feathers being broadly tipped with grey; chin, throat, breast, and sides chestnut-brown; belly white; under tail-coverts slate grey; legs and toes pale reddish-brown; claws darker.

The female resembles the male, except that she is more tinged with brown on her upper- and lighter

DARTFORD WARBLER'S NEST AND YOUNG.

on her under-parts. The chestnut-brown does not, however, extend so far down the breast.

The Dartford Warbler has the power of partly erecting the feathers on the head, so as to form a kind of crest.

*Situation and Locality.*—In the lower parts of thick furze bushes; very locally and sparingly on commons and other places covered by furze bushes, principally in the counties along the south coast of England. It was at one time not supposed to nest north of the Thames, but Mr. Dixon has proved that it does so as far north even as Yorkshire. It is not found either in Scotland or Ireland.

*Materials.*—Small and slender branches of furze, grass stalks, bits of moss and wool, with an inner lining of fine grass, and sometimes a few horsehairs. It is a somewhat slight structure, and has been likened to that of the Whitethroat.

*Eggs.*—Four or five, greenish or buffish-white in ground-colour, speckled all over with dark olive-brown, and underlying markings of grey, which generally become more dense at the larger end, and form a kind of zone. There is very little difference indeed between the eggs of this bird and those of the Whitethroat, except that the markings are more conspicuous. Size about .68 by .5 in. (*See* Plate III.)

*Time.*—April, May, and June.

*Remarks.*—Resident. Notes: *pit-et-chou-cha-ch-cha.* Local or other name: Furze Wren. Sits closely, and slips away quietly. The photograph showing adult bird with food in its bill was secured through the kindness of a friend who carefully placed meal worms day by day on the furze bush represented until the individual figured grew accustomed to finding and feeding upon them.

## WARBLER, GARDEN.

(*Sylvia hortensis.*)

Order PASSERES ; Family SYLVIIDÆ (WARBLERS).

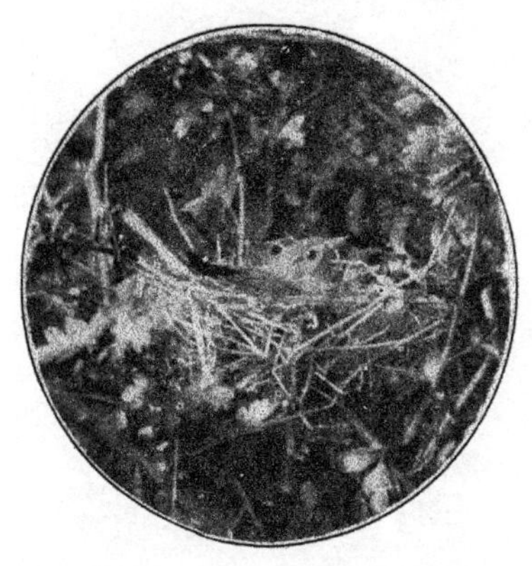

GARDEN WARBLER ON NEST.

*Description of Parent Birds.*—Length about six inches ; bill fairly long, straight, strong, and dark brown in colour. Irides hazel. Head, neck, back, wings, and tail uniform light brown, slightly tinged with olive ; chin, throat, breast, belly, vent, and under tail-coverts dull brownish-white, dark on the throat and breast, and light on the belly ; legs, toes, and claws purple-brown.

The female is similar to the male in appearance.

*Situation and Locality.*—Generally a few feet from the ground in thorn bushes, briars, brambles, gooseberry bushes, nettles, and peas. Sometimes lower down in coarse grass and taller wild plants. In woods, clumps of trees growing beside streams, shrubberies, thick hedges, orchards, and gardens, sparingly, in suitable localities nearly all over England. It also breeds in one or two parts of Wales, in the southern parts of Scotland, and in different parts of Ireland.

*Materials.*—Straws, blades of grass, fibrous roots, sometimes a little wool or moss, and lined with horsehair. It is a somewhat flimsy structure.

*Eggs.*—Four to six, generally four or five, varying in ground-colour from white to greenish-white

GARDEN WARBLER'S NEST AND EGGS.

or yellowish stone-grey, blotched, spotted, and clouded with brown of various shades ; deep olive, with underlying markings of ash-grey. The markings are variously distributed, occasionally being congregated at the larger end. Some specimens are marbled with brown. Size about .77 by .6 in. Often indistinguishable from those of the Blackcap, except by a sight of the parent birds. (*See* Plate IV.)

*Time.*—May and June.

*Remarks.*—Migratory, arriving in April and May, and departing in September or October. Notes: song, deep, harmonious, and mellow; call, *tee*. Local and other names: Pettychaps, Fauvette, Greater Pettychaps, Fig Bird. Sits closely.

---

## WARBLER, GRASSHOPPER.

(*Locustella nævia.*)

Order PASSERES ; Family SYLVIIDÆ (WARBLERS).

GRASSHOPPER WARBLER ON NEST

*Description of Parent Birds.*—Length about five and a half inches ; bill of medium length, straight, strong, and brown in colour. Irides brown. Crown, nape, back of neck, back, and wings olive-brown, the centre of each feather being of a darker tinge ; tail rather long, much rounded at the tip, and brown, barred with a paler tinge of the same colour ; chin, throat, breast, and all under-parts pale brown, darker on the flanks.

GRASSHOPPER WARBLER'S NEST AND EGGS.

The neck and breast are spotted with darkish brown; legs and toes pale brown; claws light horn colour.

The female is very similar to the male, but is said to lack breast spots.

*Situation and Locality.*—On or near the ground under furze and other small bushes, in tufts of tall rank grass growing at the foot of a hedgerow, and similar situations affording plenty of cover. In woods, on commons, fens, clumps of trees with plenty of undercover, thickets, and coppices. Pretty generally throughout England and Wales, but more sparingly distributed in Scotland and Ireland. The position of the nest and the skulking, mouse-like habits of its owner make it very difficult to find.

*Materials.*—Strong dry grass and moss, with an inner lining of finer grass. The nest is pretty deep and well built.

*Eggs.*—Four to seven, pale rosy-white, profusely spotted and speckled all over with reddish-brown. Sometimes the markings are more numerous at the larger end, and occasionally a few thin, hair-like streaks are present. Size about .72 by .54 in. (*See* Plate IV.)

*Time.*—May, June, and July.

*Remarks.*—Migratory, arriving in April and May, and departing in September. Notes: call, *tic, tic*; song, a chirping noise, similar to that made by a grasshopper, but louder and longer. Local and other names: Reeler, Cricket-Bird, Grasshopper Lark. Leaves the nest quietly and quickly, in the most mouse-like fashion, and hides in the surrounding undergrowth.

## WARBLER, MARSH.

(*Acrocephalus palustris.*)

Order PASSERES; Family SYLVIIDÆ (WARBLERS).

MARSH WARBLER ON NEST.

*Description of Parent Birds.*—This bird very closely resembles the Reed Warbler, and it is only within recent years that it has been admitted to be a distinct British breeding species. Mr. Harting has done much to establish this fact, and specimens have been seen and procured in different parts of the country. Length about five and a half inches. Bill shorter and broader than in the case of the Reed Warbler, nearly straight, dark brown above, and pale brown below. Irides hazel. Mr. Seebohm, who has had special facilities for examining specimens, describes the bird as follows, in his admirable work on British Birds:

"The Marsh Warbler has the general colour of the upper-parts varying from olive-brown in spring plumage to earthy-brown in summer, with a scarcely perceptible shade of rufous after the autumn moult, slightly paler on the rump. The eye stripe is nearly obsolete, and the innermost secondaries have broad, ill-defined pale edges. The breast, flanks, and under tail-coverts are pale buff, shading into nearly white on the chin, throat, and the centre of the belly. . . . Legs, feet, and claws horn colour."

Dresser and Sharpe say that this Warbler has the legs of a pale flesh-brown, and that those of the Reed Warbler are dark slaty-brown.

MARSH WARBLER'S NEST AND EGGS.

The female resembles the male, but is somewhat smaller in size.

*Situation and Locality.*—A celebrated Continental authority says that the nest is situated in low bushes, overgrown with nettles, reeds, and other plants, and that, unlike the Reed Warbler, which builds its nest amongst the reeds growing from the water, this bird builds its nest amongst vegetation growing from the bank of a stream or pond, and is never situated over water. The limited number of nests I have seen certainly answered this description exactly. The structure is placed from a few inches to several feet from the ground in swamps and other places affording plenty of rough undergrowth cover. It has been met with in the west of England, where our illustrations were secured in Surrey, and in the Fen country.

*Materials.*—Dry grass-stems, dead leaves, moss, and downy-fibre, with a lining of horsehair. The nest is not so deep as that of the Reed Warbler, and lacks the wool which is so often used by the last-named bird.

*Eggs.*—Four to seven, varying in ground-colour from greenish-white to greenish-blue, moderately clouded and spotted with olive-brown, and underlying markings of grey. The spots vary in size, intensity, quantity, and disposition, but are generally most numerous at the larger end of the egg. Their paler ground-colour generally distinguishes them from those of the Reed Warbler. Size about .72 by .54 in. (*See* Plate IV.)

*Time.*—June and July.

*Remarks.*—Migratory, arriving in May and departing in August. Notes: call and alarm, very similar to those of the Reed Warbler, but the song

MARSH WARBLER AND NEST.

is far finer, more melodious, and varied. It is delivered during the night as well as by day in a similar way to that of the Nightingale, and the bird is a wonderful mimic. Local or other name: none. Not a very close sitter, but wonderfully adroit in slipping off the nest and hiding in surrounding vegetation.

---

## WARBLER, REED. *Also* REED WREN.

(*Acrocephalus streperus.*)

Order PASSERES; Family SYLVIIDÆ (WARBLERS).

YOUNG REED WARBLER.

*Description of Parent Birds.*—Length about five and a half inches. Bill fairly long, strong, nearly straight, dark horn colour on the upper mandible, and lighter on the under, which is yellowish at the base. Irides light yellowish-brown. A streak of cream colour runs from the base of the beak over the eyes. Head, neck, back, wings, rump, upper tail-coverts, and tail-quills pale brown, with a tinge of chestnut, which is most pronounced on the rump; wing-quills dusky, and bordered with pale brown. Chin and throat white; breast, flanks, and under tail-coverts white, tinged with cream colour; belly white. Legs and toes dusky or slaty brown.

The female is rather smaller than the male, but similar in plumage.

*Situation and Locality.*—The nest is slung or suspended between the stems of reeds, at varying heights above the water. It is supported generally by three reeds, but upon occasion by two, four, or even five. Specimens may sometimes be met with amongst the branches of willow and other trees growing near sluggish water. I have known a nest near to Stratford-on-Avon built at some distance from the water on account of persistent persecution, there is reason to believe, by people who cut the beautiful structure and the reeds supporting it out and carry them away as ornaments. In reed beds, osier beds, and other places where suitable cover may be found, on the banks of broads, ponds, reservoirs, and sluggish streams, principally in the eastern counties. It does not breed in the extreme western peninsula of England, and is rare in the northern counties. It is not known to breed in Scotland or in Ireland, and is said to be somewhat rare in Wales.

*Materials.*—Long blades of dried grass, seed-branches of reeds, roots, dry leaves, and wool, lined with fine grass and hairs. The nest is very deep, a necessity occasioned by its supports being swayed to and fro by gusts of wind.

*Eggs.*—Four or five, dull greenish-white, greyish-green, or pale greenish-blue, spotted, blotched, and blurred with darker greyish-green and light brown. A few black spots or streaks of dark brown are sometimes present. They are variable, both in the tint of the ground-colour and markings. Their darker ground-colour and the situation of the nest distinguish them from the eggs of the Marsh Warbler. Size about .74 by .53 in. (*See* Plate IV.)

*Time.*—End of May, June, and even at the beginning of July eggs may be found.

*Remarks.*—Migratory, arriving in April and May, and leaving in September. Notes: varied, loud,

REED WARBLER'S NEST.

and hurriedly delivered. Naumann represents them as *tiri*, *tier*, *zach*, *zerr*, *scherk*, *heid*, *tret*, each note being repeated a number of times. Local and other names: Night Warbler, Reed Wren, and in some localities Reed Sparrow. Sits rather closely, and is noisy when disturbed. I have on several occasions watched the female industriously carrying piece after piece of building material to her nest whilst her mate flitted about from place to place pouring out his notes in breathless and vehement haste, but never once attempting to assist her in her work.

REED WARBLER ON NEST

---

## WARBLER, SAVI'S.

(*Locustella luscinioides.*)

Order PASSERES; Family SYLVIIDÆ (WARBLERS).

This bird used formerly to breed in the Fen country, but has long since ceased to do so on account of the drainage carried out therein, and is now only a very rare summer visitor, so that I fear we must now number it amongst Britain's lost breeding birds.

## WARBLER, SEDGE.

(*Acrocephalus phragmitis.*)

Order PASSERES; Family SYLVIIDÆ (WARBLERS).

SEDGE WARBLER.

*Description of Parent Birds.*—Length four and three-quarter inches. Bill fairly long, straight, pointed, and brown, yellowish at the base of the under mandible. Irides brown. Crown of the head streaked with light and dark brown longitudinal lines; back of neck, back, and wing-coverts light reddish-brown, mixed with a darker tint of the same colour, wing-quills dark brown, bordered with lighter tinge; rump and upper tail-coverts tawny; tail-quills brown, indistinctly barred; from the base of the beak a yellowish-white streak runs over the eye and ear-coverts, the latter of which are brown. Chin and throat white; breast, belly, and under tail-coverts pale buff; under-side of tail-quills dusky brown; flanks bright buff. Legs, toes, and claws pale brown.

The female is darker on the under-parts, and less rufous on the under tail-coverts.

*Situation and Locality.*—Amongst thick, coarse, climbing herbage, brambles, wild rose, and other bushes, near streams, rivers, and swamps, pretty generally throughout the British Isles.

*Materials.*—Grass, coarse bents, and bits of moss, sometimes none of the latter whatever, lined internally with horsehairs, and occasionally with willow down. It is a deep, cup-shaped, loosely-built structure.

*Eggs.*—Five or six, pale yellowish or umber

SEDGE WARBLER'S NEST CONTAINING A CUCKOO'S EGG.

brown, sometimes a little clouded, suffused, or mottled with darker brown, and often streaked at the larger end with a few short, hairlike, black lines. Variable, both in coloration and size. Measurements about .67 by .52 in. (*See* Plate IV.)

*Time.*—May and June.

*Remarks.*—Migratory, arriving in April and May and departing in September. Notes: call, harsh and frequently uttered. The song of the male is loud, merry, imitative, and uttered often, during the night as well as the day. Local and other names: Chat, Sedge Bird, Sedge Wren, Reed Fauvette. Sits rather closely.

---

## WARBLER, WILLOW. *Also* WILLOW WREN.

(*Phylloscopus trochilus.*).

Order PASSERES; Family SYLVIIDÆ (WARBLERS).

WILLOW WARBLER.

*Description of Parent Birds.*—Length about five inches; bill rather short, slender, slightly curved, and brown in colour; under mandible paler at the base. Irides hazel. Crown, neck, back, and upper tail-coverts dullish olive-brown, wings and tail dullish slate-brown, the feathers of the former being bordered with olive-green. A pale yellow line runs over the eye and ear-coverts; chin, throat, and breast whitish-yellow; belly, flanks, and lower tail-coverts greyish-white, slightly tinged with yellow; legs, toes, and claws light brown.

The female is very similar in all respects to the male. The bird is larger than the Chiffchaff, and the feathers in its nest readily distinguish it from the structure built by the rarer Wood Wren.

*Situation and Locality.*—On the ground amongst coarse grass and weeds, entwining themselves round slender twigs of low, open bushes growing on banks. I have found it most frequently on banks near willow- and alder-fringed streams whilst trout-fishing, and have often sat and watched the hen hop about restlessly, and after a great deal of timid hesitation, re-enter her nest. I was shown two nests in Westmorland during June, 1894, in holes in walls. One was at least three feet from the ground, and the other about a couple, not reckoning a high bank upon which the wall stood. In 1905 I found a nest in some ivy growing against a stable wall at Dingwall, at least six feet from the ground. Throughout the British Isles, wherever trees and bushes are to be found in sufficient quantities.

*Materials.*—Dead grass, moss, dead fern-fronds and leaves, lined with horsehair, cowhair, and liberal quantities of feathers. It is dome-shaped, with a hole in front which is somewhat larger than that of the Chiffchaff.

*Eggs.*—Four to eight, generally six or seven; white, spotted variably with pale rusty-red. Sometimes the spots are small and scattered pretty evenly over the surface; at others they are larger, less numerous, and more thickly congregated round the larger end. Pure white and unspotted specimens have been met with. The pale rusty-red markings distinguish the eggs of this bird from those of the Wood Wren and Chiffchaff. Size about .64 by .47 in. (*See* Plate IV.)

WILLOW WARBLER'S NEST AND EGGS.

*Time.*—April, May, June, and July, although the last-named month is somewhat late. I found some years ago a member of this species in Hertfordshire sitting on eggs as late as the first week in August.

*Remarks.*—Migratory, arriving in March and April, and departing in August and September according to some authorities, and October according to others. Specimens have been seen during the winter in the southern counties of England. Notes: long and shrill. Local and other names: Oven Bird, Jenny Wren (a name also applied to Common Wren), Scotch Wren, Yellow Wren, Hay Bird (a name also applied to the Whitethroat), Huckmuck (also applied to the Long-tailed Tit), Ground Wren, Mealy Mouth. A close sitter. It has been asserted that female Willow Warblers sometimes hover over their nests like humming birds, but, although I have watched a good many individuals of this species during the last thirty years, I have never once seen the bird practise the feat.

YOUNG WILLOW WARBLERS.

## WARBLER, WOOD. *Also* Wood Wren.

(*Phylloscopus sibilatrix.*)

Order Passeres ; Family Sylviidæ (WARBLERS).

WOOD WARBLER.

*Description of Parent Birds.*—Length about five inches; bill rather short, slender, slightly curved, and brown; crown, nape, lesser wing-coverts, back, and upper tail-coverts olive-green, tinged with yellow; wings and tail dusky, bordered with yellow of varying shades. A line of bright primrose-yellow runs from the base of the bill over the eye and ear-coverts; cheeks, chin, throat, and breast yellow; belly, vent, and under tail-coverts white; legs, toes, and claws brown. It is distinguished from the Willow Warbler by its broader yellow band over the eye, greener upper- and whiter under-parts, and longer wings.

The female is said to be a trifle larger than the male, but is similar in plumage.

*Situation and Locality.*—On the ground amongst thick herbage or dead brackens, in old plantations, woods, and other places well supplied with tall trees. Scattered generally throughout England and Wales, most numerous in some parts of Yorkshire and Durham. Met with in Scotland, but rare in Ireland.

*Materials.*—Dead grass, moss, and leaves, lined with fine grass and horsehair. Rarely feathers are said to be found, but I have never met with any. It is oval, and domed like those of the Chiffchaff and

WOOD WARBLER'S NEST AND EGGS.

Willow Wren; but is distinguished from them by having no feathers as an inner lining.

*Eggs.*—Five or six. I have generally found the former number although some authorities say the latter forms a usual clutch; white in ground-colour, thickly spotted and speckled all over with dark purplish-brown and ash-grey, most thickly at the larger end. Size about .65 by .56 in. (*See* Plate IV.)

*Time.*—May and June.

*Remarks.*—Migratory, arriving in April and departing in October. Notes: song, *twee, twee, chea, chea;* call note, *dee-ur.* Local and other names: Wood Wren, Yellow Wren. Sits closely.

---

## WATERHEN. *See* MOORHEN.

---

## WHEATEAR.

(*Saxicola œnanthe.*)

Order PASSERES; Family TURDIDÆ (THRUSHES).

MALE WHEATEAR.

*Description of Parent Birds.*—Length about six inches. Bill fairly long, strong, and black, with a few bristles at the base. Irides hazel. Crown, nape, and back bluish-grey, tinged with light-brown; rump and upper tail-coverts white. Wings nearly black, some of the feathers edged and tipped with buff. Tail-quills, upper two-thirds white, the remaining third black and broad. From the base of the beak, through the eye to the ear-coverts, is a band of black, over which is one of white, running from the forehead. Chin and throat dull white; breast pale cream colour, turning to a dull yellowish-white on the remainder of the under-parts. Legs, toes, and claws black.

The female is not nearly of such distinctive coloration, and is browner on her upper-parts.

*Situation and Locality.*—Holes in dry walls, heaps of stones, old mine hillocks, under lumps of stone jutting from steep hillsides, in chinks of rock, quarries, peat stacks, and occasionally in rabbit-burrows. On high moorland and uncultivated districts bare of trees but abounding in rocks. To be met with in suitable districts over the whole of the British Isles, but most numerous in the north of England, Wales, Scotland, Shetlands, Orkneys, and Ireland.

*Materials.*—Roots, dead grass, moss, lined with wool, hair, rabbits' down, and feathers, often loosely and clumsily put together. The materials named are, of course, not all present in the same nest, but are found according to the facilities the bird may enjoy for picking them up.

*Eggs.*—Four to seven, more generally five or six, of a pale greenish-blue, unspotted. I have found specimens nearly white sometimes, and they are said to be met with occasionally with a few small rusty spots on the larger end. Size about .83 by .61 in. (*See* Plate IV.)

*Time.*—April, May, and June.

*Remarks.*—Migratory, arriving in March and departing in August or September, stray individuals sometimes lingering as late as December. Notes:

OLD MOUNTAIN LIMEKILN IN WHICH TWO PAIRS OF WHEATEARS NESTED IN 1894.

WHEATEAR'S NEST AND EGGS.

*chick-chack-chack*. Local and other names: White Rump, Fallow Chat, Fallow Smick, Chacker, Chack-bird, Clodhopper, Fallow Finch. A close sitter.

YOUNG WHEATEARS.

---

## WHIMBREL.

(*Numenius phœopus.*)

Order Limicolæ; Family Scolopacidæ (SNIPES).

*Description of Parent Birds.*—Length about sixteen inches. Bill long, slender, curved downward, and brown; dark at the tip, and lighter towards the base. Crown dark brown, with a light central streak, and another passing from the base of the beak over the eye and ear-coverts; neck brownish-grey, streaked with dark brown. Wings dark brown, the feathers being margined and spotted with pale brown and white; rump white, streaked sparingly with brown. Tail-quills ash-brown, barred with darkish brown. Chin white; upper breast light brown, streaked with a darker tinge; lower breast and belly almost white; flanks dull white, barred with brown. Lower tail-coverts white, streaked

WHIMBREL'S NEST AND EGGS.

with brown. Legs and toes bluish-black; claws black. Distinguished from the Curlew by being considerably smaller in size and having a difference in call note.

The female is similar in plumage, but about two inches greater in length.

*Situation and Locality.*—On the ground, amongst the heather, or in the shelter of a tuft of grass, on open moors in the Shetlands, Orkneys, and Hebrides. I have only met with its nest in one place in the Shetlands, and although I see small flocks until the middle of June in the Outer Hebrides each year, have never found a nest there.

*Materials.*—A few blades of dried grass, used as a lining to the depression chosen.

*Eggs.*—Four, pear-shaped, olive-green to olive-brown in ground-colour, blotched and spotted with varying shades of brown and light grey. They resemble the darker varieties of the Common Curlew's, but are smaller, and are rather larger and more pear-shaped than those of Richardson's Skua, with which they are likely to be confused. Size about 2.35 by 1.65 in. (*See* Plate IX.)

*Time.*—May and June.

*Remarks.*—Migratory, arriving at its breeding grounds in April and May and departing in September. Notes: *tetty, tetty, tetty, tet.* Local and other names: Whimbrel Curlew, Little Whaap, Lang Whaap, Jack Curlew, Half Curlew, Curlew Knot, Maybird, Mayfowl, Titterel, Seven Whistler, Stone Curlew (a name also given to the Norfolk Plover or Thicknee), Chequer Bird. Not a close sitter, but makes a considerable outcry when disturbed.

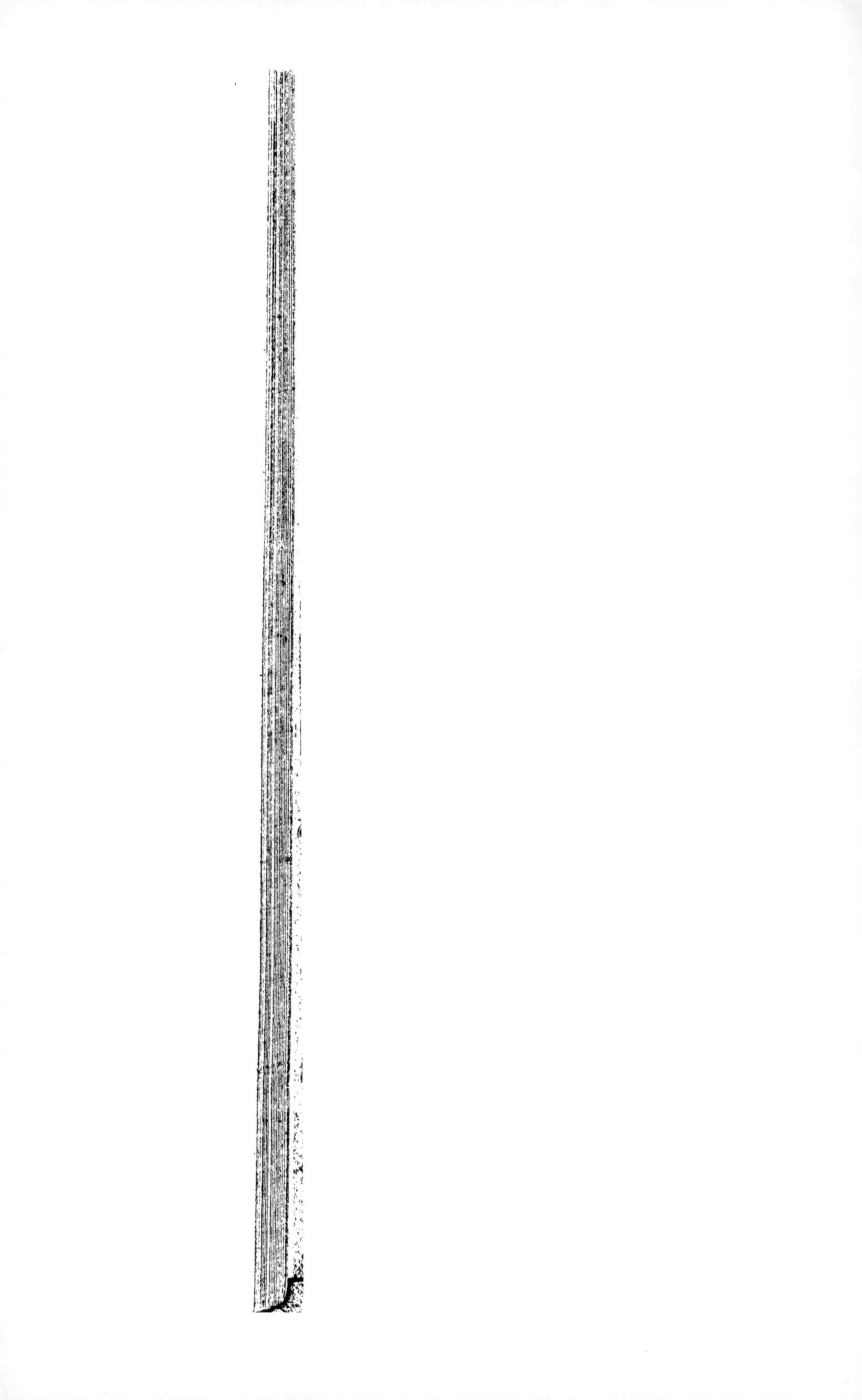

RED-LEGGED PARTRIDGE.
(*See p.* 266.)

RED GROUSE.
(*See p.* 153.)

PTARMIGAN.
(*See p.* 304.)

COMMON PARTRIDGE.
(*See p.* 264.)

PHEASANT.
(*See p.* 280.)

WATER RAIL.
(*See p.* 310.)

CAPERCAILLIE.
(*See p.* 25.)

QUAIL.
(*See p.* 309.)

BLACK GROUSE.
(*See p.* 148.)

NOTE.—*In referring to the eggs the above names should be read from left to right.*

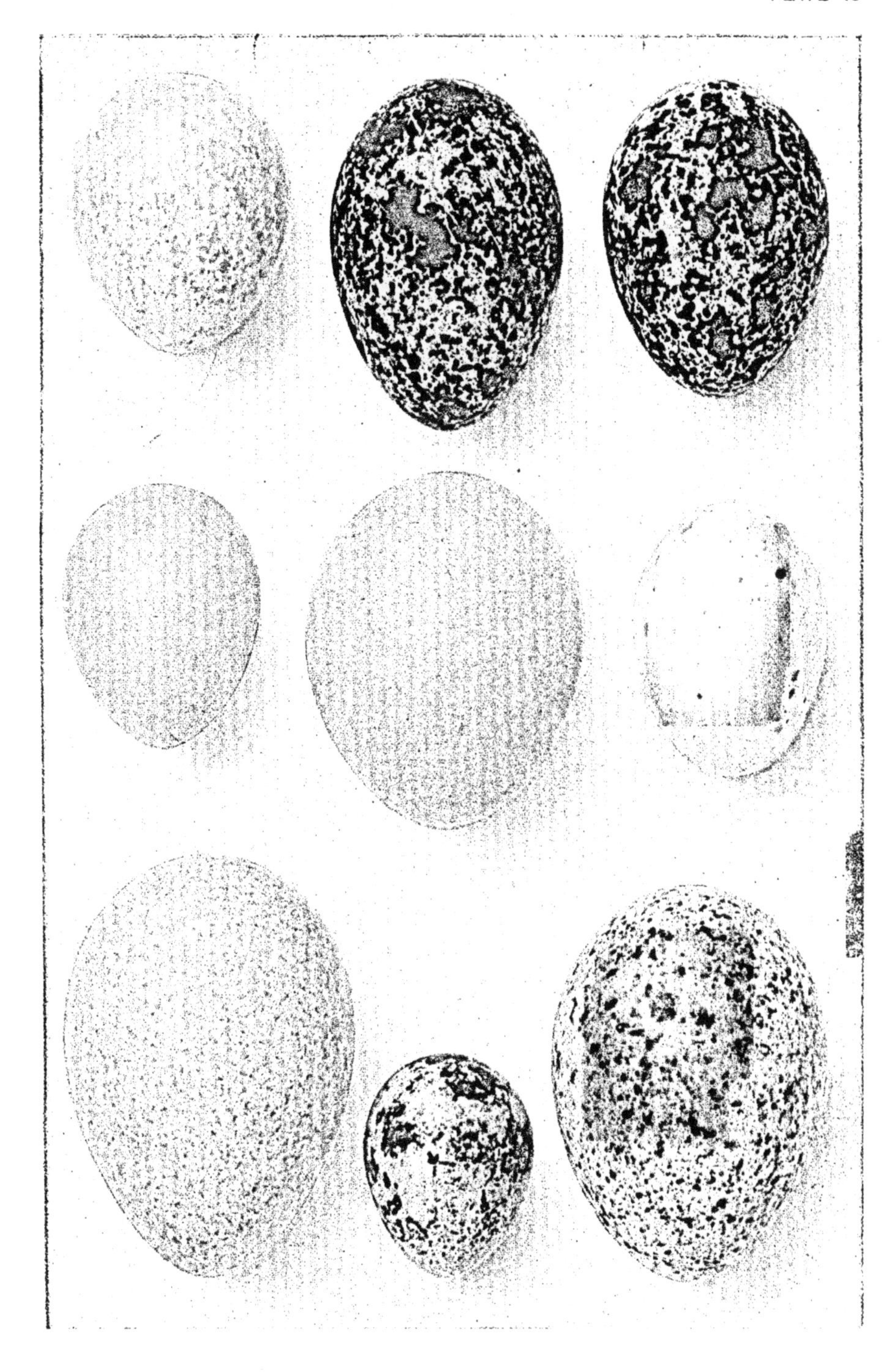

## WHINCHAT.

(*Pratincola rubetra.*)

Order PASSERES ; Family TURDIDÆ (THRUSHES).

*Description of Parent Birds.*—Length about five and a quarter inches ; bill rather short, straight, and shiny black. Irides brown. Crown, nape, back, and smaller wing-coverts of two shades of brown, the feathers being dark in the centre and light round the edges ; wings dark brown, the secondaries and tertials being edged with a paler hue ; bastard or spurious wing white ; upper half of tail white ; lower dark brown, edged with a paler tinge of the same colour. From the base of the upper mandible, over the eye and ear-coverts, is a lengthy and rather broad streak of white. From the gape to the eye buff ; chin white, extending beyond the lower margin of the ear-coverts ; throat and breast light chestnut, turning to pale buff on the belly, vent, and under tail-coverts ; legs, toes, and claws black.

In the female the white over the eye and on the wing is less distinctly marked, and the breast and belly are not so rich in coloration.

*Situation and Locality.*—On or near the ground in grass-fields, rough pasture land, on commons, at the foot of gorse bushes amongst the thick, tangling grass and dead lower branches ; also amongst heather and coarse herbage. I know several small pastures in Yorkshire where pairs nest regularly year after year. The bird is extremely local, and very wary concerning the betrayal of its nest, which is well hidden ; but I must confess I have not noticed the track or maze it has been reported to make as an approach to its nest.

WHINCHAT'S NEST AND EGGS.

*Materials.*—Dead grass and moss, with an inner lining of horsehair.

*Eggs.*—Four to six, of a beautiful greenish-blue, sparingly speckled round the larger end with minute spots of reddish-brown. They are more attenuated, of a deeper blue, and less richly and clearly marked than the eggs of the Stonechat. The appearance of the parent birds, which are not often met with close together, will, however, readily settle any doubts. Size about .76 by .57 in. (*See* Plate IV.)

*Time.*—May and June.

*Remarks.*—Migratory, arriving in April and departing in September or October. Notes: *u-tac.* Local and other names: Furze Chat, Grass Chat. Sits lightly, and, as before mentioned, is extremely wary. I detected the one figured opposite by watching through my field-glasses the female go on to her nest, and then directing my brother to the spot by signs.

---

## WHITETHROAT.

(*Sylvia cinerea.*)

Order PASSERES ; Family SYLVIIDÆ (WARBLERS).

WHITETHROAT ON NEST.

*Description of Parent Birds.*— Length about five and a half inches; bill somewhat short, straight, and brown, lighter towards the base of the under mandible. Irides yellowish. The whole of the upperparts are brown, greyish on the head and neck, and reddish on the other parts; wing-quills greyish-brown, some of the smaller ones being edged with reddish-buff; tail-quills dull brown, some of the outer ones being edged and tipped with dirty white; chin and throat white; breast, belly, and under-parts generally, pale grey, tinted with a beautiful rosy flesh colour; legs, toes, and claws brown, lightest on the first named.

WHITETHROAT'S NEST WITH NEWLY-HATCHED CHICK.

The female is less distinct in coloration, lacking the grey on the head and neck and the rosy tinge on her under-parts.

*Situation and Locality.* — Amongst nettles, brambles, thick rough grass, wild rose bushes, on hedgebanks in woods; on banks of streams, and wooded commons in nearly all suitable localities throughout the United Kingdom.

*Materials.*—Dry grass stems and hair. The nest is deep, very flimsily constructed, and loosely attached.

*Eggs.*—Four to six, of a dirty greenish-white, spotted and speckled with grey and brown. The spots are larger but not so pronounced as those of

WHITETHROAT'S NEST AND EGGS.

the Lesser Whitethroat; nor do they so often form a zone at the larger end, according to my experience. The ground-colour is not so clear a white either. Size about .72 by .55 in. (*See* Plate III.)

*Time.*—May, June, and July.

*Remarks.*—Migratory, arriving in April and departing in September or October; individual specimens have, however, been observed as late as December. Notes: song "consists of numerous agreeable strains given in rapid succession" whilst the bird is in the air. Local or other name: Nettle Creeper. Sits closely.

---

## WHITETHROAT, LESSER.

(*Sylvia curruca.*)

Order PASSERES; Family SYLVIIDÆ (WARBLERS).

LESSER WHITETHROAT ON NEST.

*Description of Parent Birds.*—Length from five to five and a quarter inches; bill rather short, yellowish-brown at the base, and nearly black towards the tip. Irides from yellowish to pearl white, according to the age of the bird. Head, neck, back, rump, and upper tail-coverts greyish-brown; wing and tail-quills dusky, edged with greyish-brown; ear-coverts dark greyish-brown; chin, throat, and under-parts greyish-white, tinged between the thighs and round the vent with red, also, in some specimens, across the breast; legs, toes, and claws leaden grey.

LESSER WHITETHROAT'S NEST AND EGGS.

The female is a trifle duller in her plumage than the male, but in all other respects is very similar. *D*iffers in size, as its name implies, from the Whitethroat, and its darker ear-coverts distinguish it.

*Situation and Locality.*—In a low, sloping hedge (our full page illustration is from a photograph of a nest in such a situation and was taken in Surrey), amongst briars, brambles, nettles, gorse and low bushes, entangled by tall coarse grass and weeds; in gardens, orchards, on commons, rough waste lands, by river banks, and the sides of small woods. Fairly plentiful in the south and east of England, but rare in the west, north, and Scotland, and met with very seldom in Ireland.

*Materials.*—*D*ead grass stalks, with an inner lining of horsehair. The whole structure is but a shallow, frail network-looking affair, that can be seen through with ease. It is sometimes tied or cemented together with old cobwebs.

*Eggs.*—Four or five; white, light creamy white, or white with the faintest suggestion of green, in ground-colour, spotted and speckled with ash-grey, greenish-brown, and umber-brown. The markings generally form a belt round the larger end. Size about .66 by .52 in. *D*istinguished by small size, clean ground-colour, and bold belt-inclining spots. (*See* Plate III.)

*Time.*—May, June, and July.

*Remarks.* — Migratory, arriving in April and leaving in September. Notes: call, *check*, repeated several times, and an incessant chattering, sometimes loud and grating, at others low and not unpleasant. Local or other name: none. Sits closely.

## WIGEON.

(*Mareca penelope.*)

Order ANSERES; Family ANATIDÆ (DUCKS).

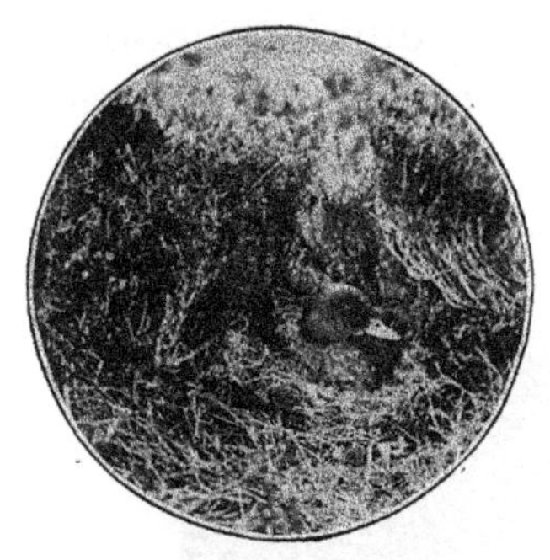

WIGEON ON NEST.

*Description of Parent Birds.*—Length about eighteen inches; bill rather short, narrow, highish at the base, and bluish-grey, tipped with black. Irides dark brown. Forehead and crown cream colour; rest of head and upper part of neck rich bay, almost black on the chin and throat. A streak of green passes backward from the eye, and the cheeks and neck are minutely spotted with blackish-green; back and scapulars greyish-white, barred with fine irregular lines of black; wing-coverts white, tipped with black; outer webs of secondaries green, tipped with black; tertials or inner secondaries black on the outer webs, which are broadly margined with white; primaries dusky-brown; upper tail-coverts freckled with grey; tail feathers long, wedge-shaped, and dusky-black, mixed with grey and brown; lower part of the neck and shoulders pale red; breast, belly, and vent white; sides and flanks marked with fine undulating lines of black; under tail-coverts rich black; legs, toes, and webs dark brown.

The female is a little smaller; her head and neck are brown, tinged with red and speckled with dark brown. The feathers on her upper-parts are dark brown, margined with light reddish-brown; breast pale brown; under-parts almost white; wings and tail somewhat similar to those of the male.

At the end of the spring and beginning of summer the male retires to some solitary swamp, and casts off his gay dress and assumes a sombre one, which he continues to wear until the autumn.

*Situation and Locality.*—On the ground, generally amongst fairly tall heather, in a clump of rushes, or amongst coarse grass, flags, reeds, and under dwarf bushes, cleverly concealed, in the neighbour-

NEWLY-HATCHED WIGEON IN NEST.

hood of lochs, swamps, tarns, or rivers, where the ground is rough and affording shelter; in suitable parts of the north of Scotland, the Orkneys and Shetlands, and probably in one or two parts of Ireland, although no nest of the species had been found up to the commencement of the twentieth century.

*Materials.*—Reeds, decayed rushes, leaves, and dry grass, with an inner lining of down from the bird's own body. The tufts are dark sooty-brown, with conspicuous white tips.

*Eggs.*—Six to twelve, generally seven or eight; creamy white, somewhat like those of the Gadwall. The locality of the nest, and the character of the

WIGEON'S NEST AND EGGS.

down tufts prevent any chance of confusion with that duck. Size about 2.2 by 1.5 in.

*Time.*—May and June.

*Remarks.*—A winter visitor, generally arriving in September or October, and departing north in March or April. Notes: a shrill whistle, sounding like "*whee you.*" Local and other names: Whew Duck, Whewer, Pandle Whew, Yellow Poll, Easterlings, Whim. A fairly close sitter, although difficult to photograph on the nest. Our initial to this article was secured by building a stone hiding house close by the nest.

---

## WOODCOCK.

(*Scolopax rusticula.*)

Order LIMICOLÆ; Family SCOLOPACIDÆ (SNIPES).

WOODCOCK ON NEST.

*Description of Parent Birds.*—Length about fifteen inches. Bill long, straight, dark brown at the tip, and pale reddish-brown at the base. Irides dark brown. A dark streak of brown extends from the gape to the eye. Head and upper-parts a mixture of rusty-brown, black, and grey, which occur in each feather and produce a handsomely variegated effect. Cheeks and the whole of the under-parts yellowish-white, numerously barred with dark wavy lines. Legs and toes brown; claws black.

The female is similar in plumage to the male, but both are subject to great variation in size and colour.

WOODCOCK'S NEST AND EGGS.

*Situation and Locality.*—On the ground, amongst dead grass, under brackens, ferns, brambles, and sometimes amongst dead leaves at the foot of a tree; in woods, plantations, coppices, and forests with plenty of the undergrowth just named; very sparingly in many localities but pretty generally throughout the United Kingdom. I have seen as many as three nests during one morning in Northumberland.

*Materials.*—Dry grass, fern-fronds, and dead leaves, placed in some natural, dry, and sheltered hollow; sometimes a suitable declivity is scraped by the bird.

*Eggs.*—Four, yellowish-white to buffish-brown, blotched with pale chestnut-brown and ash-grey. Size about 1.7 by 1.35 in. (*See* Plate IX.)

*Time.*—March, April, and May.

*Remarks.*—A winter visitor, although numbers stay with us all the year round. Notes: alarm, *skaych*, somewhat resembling that of the Snipe. Local or other name: none. Harmonises in appearance wonderfully well with its natural surroundings, and sits very closely indeed.

NEWLY-HATCHED WOODCOCK IN NEST.

## WOODLARK.

(*Alauda arborea.*)

Order Passeres ; Family Alaudidæ (LARKS).

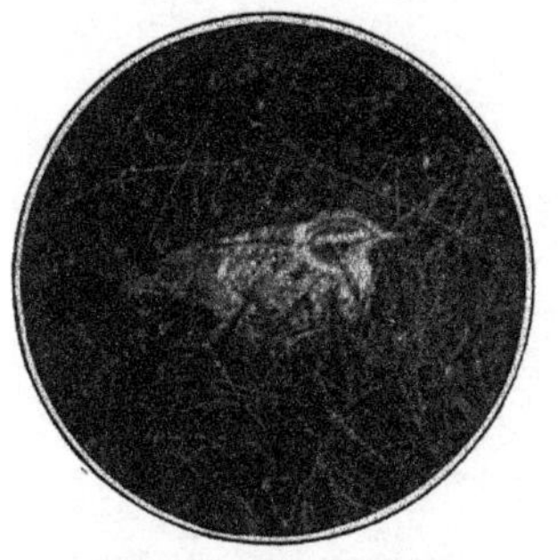

YOUNG WOODLARK.

*Description of Parent Birds.*—Length about six inches. Bill of medium length, straight, and dusky-brown, lighter at the base of the under mandible. Irides hazel. Crown light brown, streaked with a darker shade of the same colour ; feathers form an erectile crest. Over the eye and ear-coverts is a streak of pale yellowish-brown. The upper parts of the body are of a light reddish-brown, streaked and patched with dusky on the neck and back. Wing-quills dusky, bordered with brown. Tail-coverts long and brown. Tail short, outer feathers on either side brownish-black, tipped with dirty white. Throat, breast, belly, and vent yellowish-white, tinged with brown, the first-named being sparingly speckled and streaked with a darker hue ; breast streaked and spotted more thickly with the same colour. Legs, toes, and claws brown ; hind claw long and curved.

The female is smaller than the male, and is said to be yellower on the breast and to have larger markings. Distinguished from the Skylark by its more slender bill and shorter tail.

*Situation and Locality.*—On the ground, usually well concealed by a tuft of grass, heather, or low plant ; sometimes at the foot of a tree or on the side of a bank, in fields and pastures, on commons

WOODLARK'S NEST AND EGGS.

and heaths adjoining woods, copses, and plantations, most numerously in the southern and eastern counties, occasionally in the north, in Wales, and rarely in Scotland and Ireland.

*Materials.*—Coarse grass on the outside, finer grass, moss, and hair as an inner lining. The nest is placed in a little hollow, either natural or scratched out by the bird.

*Eggs.*—Four or five, pale greenish-white, light brownish-yellow, or pale reddish-white in ground-colour, thickly speckled and spotted with dull reddish-brown, and underlying markings of dark grey. The markings sometimes form a zone at the larger end. Size about .84 by .65 in. Distinguishable from those of the Skylark by small reddish-brown spots and lighter and less obscured ground-colour. (*See* Plate II.)

*Time.*—March, April, May, and June.

*Remarks.*—Resident and migratory. Notes: sings on the wing and perched on the boughs of trees; call, uttered constantly during flight, *tweedle, weedle, weedle.* Local or other name: none. A close sitter.

---

## WOODPECKER, GREATER SPOTTED.

(*Dendrocopus major.*)

Order PICARIÆ; Family PICIDÆ (WOODPECKERS).

*Description of Parent Birds.*—Length nearly nine and a half inches. Beak of medium length, straight, sharp at the tip, and dusky. Irides red. Forehead buffish; round the eyes and ear-coverts dirty white. Crown black; back of head bright scarlet. A black stripe commences at the gape and, widening, passes backward under the eye and ear-coverts to the nape; another commences on the side of the throat and, passing backwards, also meets the black on the back of the neck. A horizontal, elongated patch of white is enclosed by the black on either side of the neck; the back, rump, and upper tail-coverts are black. Wings black, variegated with white spots and a large patch of the same colour on the scapulars. Tail longest in the centre, and black, the outside feathers being tipped in increasing lengths from the middle with white. Chin, throat, breast, and belly dirty white; vent and under tail-coverts bright scarlet. Legs, toes, and claws greenish-grey.

The female is a little smaller than the male, and lacks the scarlet on the back of her head. Easily distinguished from the Lesser Spotted Woodpecker by its much greater size and lack of white on the back.

*Situation and Locality.*—In holes in trees, either dug by the bird's own exertions, or a decayed hole in the trunk or a branch, adapted and enlarged. It is somewhat similar to that of the other Woodpeckers, and varies from ten to twenty inches in depth. In forests, well-timbered parks, woods, and other places where old trees exist. It is found in nearly all the counties of England and Wales, excepting those north of Yorkshire, where it is a rare bird; it is local in Scotland, but apparently increasing in central and south-eastern parts; rare in Ireland.

*Materials.*—None, the eggs being laid on the powdered wood and chips produced in making the cavity for their reception.

*Eggs.*—Four to seven, occasionally as many as eight, white, unspotted, and glossy. Size about 1.05 by .75 in. Distinguished by their size and characteristics of parent birds.

SITE OF A GREATER SPOTTED WOODPECKER'S NEST.

*Time.*—May and June.

*Remarks.*—Resident, but its numbers are said to be increased in winter by Continental visitors. Notes: *gich-gich, quet-quet, tra, tra, tra.* Local and other names: Witwall, Woodnacker, Wood-pie, French-pie, Great Black and White Woodpecker, Spickel-pied Woodpecker. A very close sitter.

---

## WOODPECKER, GREEN.

(*Gecinus viridis.*)

Order PICARIÆ; Family PICIDÆ (WOODPECKERS).

GREEN WOODPECKER.

*Description of Parent Birds.*—Length about thirteen inches. Beak rather long, strong, and dusky in colour. Irides greyish-white. Crown, crimson; neck, back, lesser wing-coverts, and scapulars green; rump yellow; upper tail-coverts yellow, tinged with green in parts. Wing-quills dusky, barred and spotted with buffish-white, some of the lesser being margined with olive-green. Tail-quills dusky, barred with greyish-brown, some of them being margined with green. From the base of the beak, round and behind the eyes, the feathers are black. A crimson streak, bordered with black, runs from the gape some little way down the sides of the neck. Chin, throat, breast, and all under-parts pale greyish-green. Legs, toes, and claws ash-colour. The toes are disposed, two in front and two behind, and claws hooked.

The female has less crimson on her crown, and

GREEN WOODPECKER'S NESTING-HOLE

none at all from the gape down the sides of the neck, which is black.

*Situation and Locality.*—In holes in trees, generally dug by the bird's own exertions, those composed of soft wood being preferred. The hole is from ten to eighteen inches deep. It breeds in suitably wooded localities nearly all over England and Wales, but is least numerous in the northern counties, and does not breed in either Scotland or Ireland.

*Materials.*—Only the chips and bits of decayed wood that have become detached in hewing the nesting-place.

*Eggs.*—Five to seven, sometimes eight, pure white, unspotted, and glossy. Size about 1.3 by .92 in. Distinguished by large size and appearance of parent birds.

*Time.*—April and May.

*Remarks.*—Resident. Notes, several, which have been represented as follows: male spring note, *tiacacan, tiacacan;* call, used all the year round, *pleu, pleu, pleu.* Some writers represent the call as *yaffa, yaffa, yaffle!* Local and other names: Rainbird, Popinjay, Awlbird, Yaffle, Tongue Bird, Gally, Rain-fowl, Pick-a-tree, Whetile, Woodwale, Wood-speight, Yaffingale. A close sitter, often occupying the same nest year after year, when not evicted by Starlings. I have known a Green Woodpecker occupy herself day by day for a whole fortnight in laboriously digging out a nesting hole, and directly it was completed she was turned out and the place taken possession of by a pair of old Starlings.

## WOODPECKER, LESSER SPOTTED.

(*Dendrocopus minor.*)

Order PICARIÆ ; Family PICIDÆ (WOODPECKERS).

*Description of Parent Birds.*—Length about five and three-quarter inches. Beak of medium length, broad at the base, straight, and leaden grey in colour. Irides hazel. Crown bright scarlet, sides of head brownish-white. A black stripe runs from the base of the beak over the eye to the nape, which is black also ; another runs from the base of the under mandible below the eye, and beneath the ear-coverts. Back of neck and upper back black. Wings black, barred and spotted with white ; middle of back white, barred with black ; rump and upper tail-coverts black. Tail-quills black, some of them edged and tipped with white, others white, barred with black. Chin, throat, and under-parts brownish-white ; sides of breast and flanks streaked and slightly barred with black. Legs, toes, and claws lead grey.

The female has the crown brownish-white, occasionally shaded with red ; black of nape commences further forward, and under-parts are darker.

*Situation and Locality.* — In a hole, dug or enlarged, in the stem or large branch of a tree. Pear and apple trees appear to be favourites. The hole is usually from six or seven to twelve or fourteen inches deep. In orchards, spinneys, parks, woods, and well-timbered districts generally, in the south and west of England, and as far north as York.

*Materials.*—None, the eggs being laid on the wood-dust and fine chips at the bottom of the hole.

*Eggs.*—Five to nine, six being the general number, white and glossy. Size about .76 by .58 in.

LESSER SPOTTED WOODPECKER'S NESTING SITE.

Distinguished from those of the Wryneck only by smaller size and a sight of parents.

*Time.*—April and May.

*Remarks.*—Resident. Notes: *tic-tic*, or *kink-kink*. Local and other names: Crank Bird, Pump-borer, Least-spotted Woodpecker, Little Black and White Woodpecker, Hickwall, Barred Woodpecker. Sits closely.

---

## WREN, COMMON.

(*Troglodytes parvulus.*)

Order PASSERES; Family TROGLODYTIDÆ (WRENS).

COMMON WREN'S NEST.

*Description of Parent Birds.*—Length a little under four inches; bill of moderate length, slightly curved downwards, dark brown on the top, and light brown underneath. Irides hazel. A dull, whitish line runs over the eye and ear-coverts. Head, nape, back, and rump reddish-brown, faintly marked on the two latter with wavy bars of light and dark brown; wing and tail feathers marked with light reddish-brown and black bars; chin, cheeks, throat, and breast greyish-white, tinged with buff; belly, sides, and thighs light brown, marked with narrow wavy bars of a darker tint of the same colour; under tail-coverts indistinctly spotted with black and dirty white; legs, toes, and claws pale brown.

The female is rather smaller, duller on her upper-parts, and darker beneath.

*Situation and Locality.*—I have found the nest of this bird in crevices and holes of rock, in the gnarled roots of trees growing on the banks of streams, amongst fern roots growing on the sides of banks, in banks of rivers, amongst ivy growing against walls and trees, in holes in brick and stone walls, faggot and hay-stacks; in the thatch of barns; amongst wreckage lodged in a thorn bush by a high flood; between the stems of two trees growing close together over a stream, and lodged amongst a few slender twigs sprouting from the trunk of a tree where a large branch had been lopped off, fixed to the bare trunk of a tree, and in various other situations. The nest has been found in all sorts of odd places, such as in cabbages run to seed, bodies of scarecrows, and the skeleton of a Carrion Crow hanging against a wall. Generally throughout the British Isles.

*Materials.*—Moss, dead leaves, fern-fronds, roots, dry grass-stalks, the stems of leaves, lined with hair and feathers. I remember once finding one lined entirely with the feathers of a hen grouse, whose skeleton I discovered not far away. The bird seems to have some idea of the value of harmonisation. The nest I mentioned finding amongst fern roots was composed outside of dead fern-fronds, and the one between the stems of two trees growing close together over a brook, of bright green moss, matching that upon the trunk on either side exactly. The one in the twigs growing from the stem of a tree was composed outside entirely of dead leaves, and looked exactly like an accidental collection. However, there are

COMMON WREN ABOUT TO ENTER NEST BUILT IN THE SIDE OF A STUMP.

exceptions to this rule, and I have found specimens made of bright green moss situated in the bleached side of a hayrick. The bird practises the curious habit of building the outer structure of several nests, but whenever one is found with an inner lining of feathers it is sure, in the absence of accidents, to be laid in.

*Eggs.*—Four to eight, generally from five to seven, although as many as twelve and fourteen,

COMMON WREN PEEPING OUT OF NEST.

and even twenty, have upon rare occasions been found. White, sparingly spotted with brownish-red of varying shades, generally distributed at the larger end. Specimens are sometimes found quite unspotted. One day I discovered a nest which contained three unspotted eggs, and
I remarked to my brother that the layer
in a poor state of health. We photo-
ed it on the Saturday afternoon, and I
he nest at six o'clock on the following
orning, and found the hen in-
t away for an hour, and when I

returned she was still on her eggs, and I discovered that she was quite dead and in a very emaciated condition. On dissection there was no sign of further eggs in her body. Size about .7 by .51 in. (*See* Plate III.)

*Time.*—April, May, June, and July.

*Remarks.*—Resident. Notes: alarm, a jarring kind of note; song, loud, joyous, and heard all the year round. Local and other names: Cutty Wren, Titty Wren, Jenny Wren (a name also applied to the Chiffchaff and Willow Wren), Tom Tit (in the North of England), Kitty Wren. A close sitter.

---

## WREN, ST. KILDA.

(*Anorthura hirtensis.*)

Order PASSERES; Family TROGLODYTIDÆ (WRENS).

YOUNG ST. KILDA WREN.

To Mr. Charles Dixon belongs the credit of first discovering the differences between this bird and its mainland representative. Whilst staying at St. Kilda in 1896 my brother and I gave the species particular attention, and our observations resulted in establishing the following facts. It is larger than its mainland representative; its beak, legs, toes and claws are a trifle stronger and lighter in colour, and its plumage is generally much paler and more distinctly marked. The bird does not cock its tail at the acute angle so characteristic of the mainland species. Sings oftener on the win-

ST. KILDA WREN'S NEST.

and builds a somewhat larger and rougher nest. It lays larger eggs, as will be seen by reference to our illustration below, and not so many of them; six appearing to be the maximum number of a clutch. (*See* Plate III.)

It is said that the Shetland Wren is also larger than its typical mainland representative, but whether it differs as widely as the bird found in St. Kilda I am unable to say.

Whether the St. Kilda Wren does or does not differ sufficiently to entitle it to a specific name I must leave systematists to decide, but there can be no doubt whatever that the publication of its points of difference was a bad thing for the bird, and it has been found necessary to protect it and its eggs by law. It is very gratifying to e able to e from onal dge e

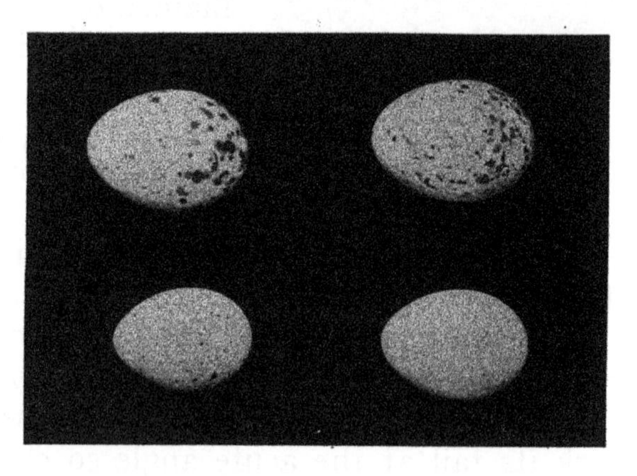
UPPER TWO EGGS LAID BY ST. KILDA WREN.
LOWER TWO BY MAINLAND WREN.

McLeod of McLeod, and his factor, Mr. John Mackenzie, are doing everything in their power to prevent its extermination by the natives and the temptation-offering collectors.

---

## WREN, GOLDEN-CRESTED. *See* Gold-crest.

---

## WREN, REED. *See* Warbler, Reed.

---

## WREN, WILLOW. *See* Warbler, Willow.

---

## WREN, WOOD. *See* Warbler, Wood.

---

## WRYNECK.

(*Iynx torquilla.*)

Order Picariæ ; Family Picidæ (WOODPECKERS).

*Description of Parent Birds.*—Length about seven inches. Beak rather short, straight-pointed, and brown. The whole of the upper-parts of the body consist of varying shades of brown, mixed with grey, pencilled, mottled, barred, and streaked with buff, greyish-white, brownish-black, and black. The top of the head is barred with blackish-brown, the nape of the neck striped with the same, and also parts of the back and wings. The wing-quills are dark brown, barred and spotted with two shades of buff. Tail-quills greyish-brown, marked with several irregular blackish-brown bars. All the under-parts are dull white, tinged with yellowish-buff on the chin, throat, flanks, and under tail-coverts, which are barred with dark brown. The breast and belly are marked with small

WRYNECK AT NESTING HOLE IN FENCE POST.

triangular spots of dark brown. Legs, toes, and claws brown.

The female is somewhat duller in her coloration.

*Situation and Locality.*—In holes in trees and posts, at varying heights from the ground and at differing depths. The deserted hole of a Woodpecker is a favourite site. In open woodlands, parks, trees growing by brooks, roads, and in fields; in the south and east of England. It is scarce in the west and north, and more so in Scotland and Ireland.

*Materials.*—Generally the decayed and powdered wood at the bottom of the hole selected. Occasionally it is said to contain other materials, such as moss, wool, hair, or feathers, but these might have been previously deposited by some other bird, as the Wryneck is not averse to using a hole so furnished. My friend, Mrs. Patterson, of Limpsfield, tells me that Wrynecks sometimes turn even the Great Tit out of a nesting box in her grounds, and after getting rid of the eggs of the rightful owner, lay in the nest.

*Eggs.*—Six to ten, generally seven or eight, pure white, unspotted, and often mistaken for those of the Lesser Woodpecker, from which they differ, however, in being a trifle larger. A sight of the parent birds is the only certain method of identification. Average size about .85 by .63 in.

*Time.*—May and June.

*Remarks.*—Migratory, arriving in April, and leaving in September. Note: a betraying *peel, peel, peel,* uttered about nine times in unbroken succession. Local and other names: Snake Bird, Cuckoo's Mate, Tongue Bird, Emmet Hunter, Long Tongue, Barking Bird. Sits closely, and hisses.

## YELLOW HAMMER. *Also* YELLOW BUNTING.

(*Emberiza citrinula.*)

Order PASSERES; Family EMBERIZIDÆ (BUNTINGS).

YELLOW HAMMER ON NEST.

*Description of Parent Birds.*—Length about seven inches. Bill short, strong, and bluish horn colour, tinged with brown. Irides dark brown. There are a few short bristles round the base of the bill. Head and nape yellow, tinged with green, and marked on the crown with a few streaks of dusky-black and olive. Back bright reddish-brown, tinged with yellowish-green. Wing-quills dusky, bordered with greenish-yellow; rump bright chestnut; tail slightly forked, dusky-black, edged with greenish-yellow, the two outer feathers being marked with white spots. Throat, breast, belly, and under tail-coverts bright yellow, the breast being sometimes marked with reddish spots, and the sides streaked. Legs, toes, and claws yellowish-brown.

The female is a trifle smaller, and much duller in her plumage. She is less yellow, and more thickly marked with brown. Her tail is also lighter, and has less white on the outsides. I have often been struck by her close harmonisation with surrounding objects when her nest is on the ground. Both sexes are subject to variation, but lack the black of the male, and the yellowish-brown of the female Cirl Bunting under the chin.

*Situation and Locality.*—On or near the ground,

YELLOW HAMMER'S NEST AND EGGS.

although specimens may sometimes be found at a height of eight or ten feet. In hedgebanks, amongst brambles, nettles, and coarse grass at the foot of light open bushes. On pieces of waste land, commons, pastures, grass-fields, and arable lands, in all suitable localities throughout the United Kingdom.

*Materials.*—Dry grass, roots, and moss, with an inner lining of fine grass and horsehair. The nest varies in bulk according to situation.

*Eggs.*—Three to six, generally four or five. Ground-colour dingy-white, tinged with purple, and streaked, veined, spotted, and blotched with dark purplish-brown, the streaks and lines generally terminating in a spot of the same colour. There are also underlying markings of purplish-grey. The purple tinge of the ground-colour and the thick scribbling lines distinguish them from those of the Cirl Bunting, with which they are likely to be confused. Subject to great variation. Size about .88 by .65 in. (*See* Plate II.)

*Time.*—April, May, June, July, and August.

*Remarks.*—Resident. Notes: *chit, chit,* followed by a long, harsh *chire-r-r.* Bechstein represents the song by *te, te, te, te, te, te, tywee,* but it is popularly interpreted in this country as *Bit o' bread and no chee-e-e-se.* Local and other names: Yellow Bunting, Yellow Yowley, Goldspink, Yoist, Yellow Yite, Yellow Yoldring, Yeldrock, Yellow Yeldring. A very close sitter.

---

PRINTED BY CASSELL & COMPANY, LIMITED, LA BELLE SAUVAGE, LONDON, E.C.

Lightning Source UK Ltd.
Milton Keynes UK
UKHW020714211218
334381UK00012B/675/P